ERGEBNISSE DER MATHEMATIK
UND IHRER GRENZGEBIETE

UNTER MITWIRKUNG DER SCHRIFTLEITUNG DES
„ZENTRALBLATT FÜR MATHEMATIK"

HERAUSGEGEBEN VON

L.V. AHLFORS . R. BAER · R. COURANT · J. L. DOOB · S. EILENBERG
P. R. HALMOS · M. KNESER · T. NAKAYAMA · H. RADEMACHER
F. K. SCHMIDT · B. SEGRE · E. SPERNER

NEUE FOLGE · BAND 19

REIHE:

MODERNE
FUNKTIONENTHEORIE

BESORGT

VON

L. V. AHLFORS

SPRINGER-VERLAG
BERLIN HEIDELBERG GMBH 1964

POLYNOMIAL EXPANSIONS
OF ANALYTIC FUNCTIONS

BY

RALPH P. BOAS, JR., AND R. CREIGHTON BUCK

SECOND PRINTING CORRECTED

WITH 16 FIGURES

SPRINGER-VERLAG
BERLIN HEIDELBERG GMBH 1964

ISBN 978-3-662-23179-1   ISBN 978-3-662-25170-6 (eBook)
DOI 10.1007/978-3-662-25170-6

© by Springer-Verlag Berlin Heidelberg 1964
Originally published by Springer-Verlag 1964
Softcover reprint of the hardcover 2nd edition 1964
Library of Congress Catalog Card Number 63-23263

Preface

This monograph deals with the expansion properties, in the complex domain, of sets of polynomials which are defined by generating relations. It thus represents a synthesis of two branches of analysis which have been developing almost independently. On the one hand there has grown up a body of results dealing with the more or less formal properties of sets of polynomials which possess simple generating relations. Much of this material is summarized in the Bateman compendia (ERDÉLYI [1], vol. III, chap. 19) and in TRUESDELL [1]. On the other hand, a problem of fundamental interest in classical analysis is to study the representability of an analytic function $f(z)$ as a series $\sum c_n p_n(z)$, where $\{p_n\}$ is a prescribed sequence of functions, and the connections between the function f and the coefficients c_n. BIEBERBACH's monograph *Analytische Fortsetzung* (Ergebnisse der Mathematik, new series, no. 3) can be regarded as a study of this problem for the special choice $p_n(z) = z^n$, and illustrates the depth and detail which such a specialization allows. However, the wealth of available information about other sets of polynomials has seldom been put to work in this connection (the application of generating relations to expansion of functions is not even mentioned in the Bateman compendia).

At the other extreme, J. M. WHITTAKER and his students have obtained many results about expansions of analytic functions in the so-called basic series which can be associated with very general sets of polynomials. (See especially WHITTAKER [2].) In that theory the degree of generality is so great that broad rather than refined results are to be expected, and the internal structure of the prescribed sequence of polynomials does not play much of a role. For example, the basic tool of the Whittaker theory is the rearrangement of power series, and so the theory is dominated by the presence of circular regions of convergence.

We have adopted an intermediate position by discussing the expansion of analytic functions in series of polynomials defined by a rather general kind of generating relation. Here we have tried to bring about a certain amount of order and completeness and to formulate results and methods in a fashion which will make them more generally accessible. While we do not obtain as much information about the expansions as is obtainable, for example, about power series, we obtain considerably more than is obtainable for more general sets of polynomials: thus we are not restricted to circular regions, or to the basic series, and we can discuss summability as well as convergence. Results ot this kind appear here and there in the literature (mostly for the classical orthogonal polynomials), but as isolated observations and not as part of a coherent theory. (An exception is MARTIN [1], where a theory of the expansion of entire functions is developed by the method which we follow in general.) The chief tool that we use is the method of kernel expansion: this is at least as old as CAUCHY's deduction of TAYLOR's theorem from CAUCHY's integral formula[1].

While little that we have to say is new in principle, some of the general theory that we present is rather more general than anything we have seen elsewhere, and some of the details seem not to have been worked out before. In particular, we have illustrated the general theory by applying it to many of the almost innumerable sets of polynomials which have been introduced into the literature for one reason or another. The material of § 8 on the possible form of multiple expansions of a given function has not been published before[2]. Some open questions are mentioned on pp. 10, 18, 27 and 29.

This study has been developed at intervals during the past twelve years. For financial support during parts of this period, we are indebted to the John Simon Guggenheim Memorial Foundation and Northwestern University (BOAS) and to the Office of Ordnance Research (BUCK).

Evanston (Illinois) June 1957   R. P. BOAS, JR.
Madison (Wisconsin) June 1957   R. C. BUCK

Preface to the revised edition

The present revision has been limited to the correction of a number of printing errors and oversights. We have not been able to include an account of the interesting recent work of FALGAS[3] which subsumes both WHITTAKER's approach and ours in a very general theory.

Evanston (Illinois) January 1964   R. P. BOAS, JR.
Madison (Wisconsin) January 1964   R. C. BUCK

---

[1] See PRINGSHEIM [1].

[2] Some of our results, as far as they concern Appell sets, were obtained independently by J. STEINBERG [1].

[3] FALGAS [4], [5].

Contents

Chapter I. Introduction 1

§ 1. Generalities 1

§ 2. Representation formulas with a kernel 4

§ 3. The method of kernel expansion 10

§ 4. Lidstone series 13

§ 5. A set of Laguerre polynomials 16

§ 6. Generalized Appell polynomials 17

Chapter II. Representation of entire functions 21

§ 7. General theory 21

§ 8. Multiple expansions 24

§ 9. Appell polynomials 28

(i)   Bernoulli polynomials and generalizations 29

(ii)   A set of Laguerre polynomials 31

(iii) Hermite polynomials 31

(iv)   Reversed Laguerre polynomials 32

(v)   Reversed Rainville polynomials 32

§ 10. Sheffer polynomials 33

(vi)   General difference polynomials 34

(vii)   Poisson-Charlier, Narumi and Boole polynomials 37

(viii) Mittag-Leffler polynomials 38

(ix)   Abel interpolation series 38

(x)   Laguerre polynomials 40

(xi)   Angelescu polynomials 41

(xii)   Denisyuk polynomials 41

(xiii) Squared Hermite polynomials 41

(xiv)   Adhoc polynomials 41

(xv)   Actuarial polynomials 42

§ 11. More general polynomials 42

(xvi)   Special hypergeometric polynomials 43

(xvii)   Reversed Bessel polynomials 43

(xviii) q-difference polynomials 44

(xix)   Reversed Hermite polynomials 45

(xx)   Rainville polynomials 46

§ 12. Polynomials not in generalized Appell form 46

Chapter III. Representation of functions that are regular at the origin 47

§ 13. Integral representations 47

§ 14. Brenke polynomials 51

(i) Polynomials generated by $A(w)(1-zw)^{-\lambda}$ 52

(ii) q-difference polynomials 54

§ 15. More general polynomials 55

§ 16. Polynomials generated by $A(w)\left(1 - zg(w)\right)^{-\lambda}$ 57
    (iii) Taylor series 57
    (iv) Lerch polynomials 57
    (v) Gegenbauer polynomials 58
    (vi) Chebyshev polynomials 58
    (vii) Humbert polynomials 58
    (viii) Faber polynomials 59
§ 17. Special hypergeometric polynomials 60
    (ix) Jacobi polynomials 60
§ 18. Polynomials not in generalized Appell form 61

Chapter IV. Applications 65
§ 19. Uniqueness theorems 65
§ 20. Functional equations 67

Bibliography 71
Index 75

Chapter I

Introduction

§ 1. Generalities

The place of our work in the theory of polynomial expansions will be seen best if we begin with some general remarks. Let $\mathfrak{P}$ be the complex linear space of all polynomials, with the topology of uniform convergence on all compact subsets of a simply-connected region Ω. The completion of $\mathfrak{P}$ is then the space $\mathfrak{A}(\Omega)$ of all functions f which are analytic in Ω. Let $\sigma = \{p_n\}$ be a sequence of polynomials which forms a basis for $\mathfrak{P}$: that is, any $p \in \mathfrak{P}$ has a unique representation as a finite sum $p = \sum c_n p_n$. It is customary to call such a σ a basic set of polynomials. Then every $f \in \mathfrak{A}(\Omega)$ is the limit of a sequence of finite sums of the form $\sum_n a_{k,n} p_n$.

Of course this by no means implies that there are numbers c_n such that $f = \Sigma c_n p_n$ with a convergent or even a summable series. One way of attaching a series to a given function is as follows. Since σ is a basis, in particular there is a row-finite infinite matrix, unique among all such matrices, such that

$$z^k = \sum_{n=0}^{\infty} \pi_{k,n} p_n(z), \qquad k = 0, 1, 2, \dots . \tag{1.1}$$

Suppose that Ω contains the origin, let f be analytic at the origin, and write

$$f(z) = \sum_{k=0}^{\infty} f^{(k)}(0)\, z^k/k! . \tag{1.2}$$

If we formally substitute (1.1) into (1.2), we obtain

$$f(z) = \sum_{k=0}^{\infty} \frac{f^{(k)}(0)}{k!} \sum_{n=0}^{\infty} \pi_{k,n}\, p_n(z) ,$$

or

$$f(z) = \sum_{n=0}^{\infty} c_n p_n(z) , \tag{1.3}$$

where

$$c_n = \sum_{k=0}^{\infty} \pi_{k,n} f^{(k)}(0)/k! . \tag{1.4}$$

The expansion (1.3) with coefficients (1.4) is the so-called basic series, introduced by J. M. WHITTAKER [1], and studied in more detail by him in a recent monograph [2] and by his students in a long series of papers[1]. A typical theorem is the following: with the basic set $\{p_n\}$ we associate two numbers ω ("order") and γ ("type"); then every entire function of growth less than order $1/\omega$, type $1/\gamma$ is represented by its basic series (1.3) with coefficients (1.4); and in general functions of more rapid growth are not so representable.

The present study arose from the observation that this theorem, while it tells nothing but the truth, does not tell the whole truth. The following simple example will serve as an illustration.

Consider the basic set defined by

$$\left.\begin{aligned}p_0 &= -1 \\ p_n(z) &= \frac{z^{n-1}}{(n-1)!} - \frac{z^n}{n!}, \quad n = 1, 2, \dots.\end{aligned}\right\} \qquad (1.5)$$

It can easily be calculated that this set is of order 1 and type 1, so that every entire function of exponential type less than 1 is represented by the basic series. As WHITTAKER's general theory predicts, this result is to be regarded as sharp. For, in this special case formula (1.4) for the coefficients takes the form

$$c_n = -\sum_{k=n}^{\infty} f^{(k)}(0). \qquad (1.6)$$

Applying this to $f(z) = e^z$, an entire function of exponential type 1, we obtain $c_n = -(1 + 1 + \cdots)$. Since the formula (1.4) thus fails to define coefficients for the basic series (1.3), it is customary to say that the basic series does not exist, and certainly does not represent e^z. However, it is easy to see that every entire function f, and indeed every function analytic at 0, has a convergent representation

$$f(z) = \sum c_n' p_n(z),$$

where $c_0' = 0$ and

$$c_n' = \sum_{k=0}^{n-1} f^{(k)}(0), \quad n = 1, 2, \dots. \qquad (1.7)$$

Moreover, since $\sum_{n=0}^{\infty} p_n(z)$ converges for every z to the sum 0, we also have $f(z) = \sum (c_n' + a) p_n(z)$ for every a, so that every function analytic at 0 has an infinity of convergent expansions in terms of the $p_n(z)$. In the light of such examples, we have felt that it is advisable to re-examine the subject.

Let us again consider the space $\mathfrak{P}$, its completion $\mathfrak{A}(\Omega)$ and a given basis σ. By $\mathfrak{E}(\sigma)$ we shall mean the subspace of $\mathfrak{A}$ consisting of all f that can be expressed in the form of a convergent series

$$f(z) = \sum_{n=0}^{\infty} c_n p_n(z), \qquad (1.8)$$

---

[1] See Math. Reviews passim, particularly under Doss, EWEIDA, MAKAR, MIKHAIL, MURSI, NASSIF, TANTAOUI; also NEWNS [1], FALGAS [1].

irrespective of the mode of formation of the c_n. We call $\mathfrak{E}(\sigma)$ the *expansion class* for σ. If $\mathfrak{E}(\sigma) = \mathfrak{A}$ and if in addition the expansion (1.8) is unique, $\{p_n\}$ is a base for $\mathfrak{A}$ and the c_n may be given the form $c_n = \mathscr{L}_n(f)$, where $\{\mathscr{L}_n\}$ is a sequence of linear functionals orthogonal to $\{p_n\}$. If $\mathfrak{E}(\sigma) = \mathfrak{A}$ and the expansion (1.8) is not necessarily unique, we call $\{p_n\}$ a *semibase*. In this case we have to make precise the notion of an expansion formula[1].

Let $\mathfrak{S}$ be the space of complex sequences $c = (c_0, c_1, \ldots)$ such that $\sum c_n p_n$ converges to an element of $\mathfrak{A}$. Let U be the linear transformation from $\mathfrak{S}$ onto $\mathfrak{E}(\sigma)$ sending $c \in \mathfrak{S}$ into $U(c) = f = \sum c_n p_n \in \mathfrak{E}(\sigma)$. Let T be any linear transformation whose domain is at least $\mathfrak{P}$ and whose range lies in $\mathfrak{S}$; suppose further that, for every $p \in \mathfrak{P}$, $UT(p) = p$, so that T is a right inverse for U on $\mathfrak{P}$. Then, in the domain of T, we have $UT(f) = f$. Denote the class of such f by $\mathfrak{E}(\sigma, T)$. Then we say that T defines an *expansion formula* applicable to the class $\mathfrak{E}(\sigma, T)$. If we represent T as a sequence of linear functionals $T = (\mathscr{L}_0, \mathscr{L}_1, \ldots)$, so that $T(f) = c$ where $c_n = \mathscr{L}_n(f)$, then for all $f \in \mathfrak{E}(\sigma, T)$ we have $f = UT(f) = U(c) = \sum \mathscr{L}_n(f) p_n$. We have thus shown that each right inverse of U on $\mathfrak{P}$ gives rise to an expansion formula and a class to which it is applicable. If U is one-to-one, there is essentially only one right inverse; in this case σ is a base for $\mathfrak{A}$. If U is not one-to-one, U has an infinite number of right inverses; among these, one may be singled out as the principal right inverse, as follows. Since σ is a basis for $\mathfrak{P}$ we may choose T so that $T(p) = (c_0, c_1, \ldots)$, the unique sequence of coefficients for which $p = \sum c_n p_n$ and $c_n = 0$ for all large n. The expansion formula corresponding to this choice of T is the basic series.

We may illustrate this with our example (1.5). Here (1.6) defines the sequence $\{\mathscr{L}_n\}$ of functionals given by

$$\mathscr{L}_n(f) = -\sum_{k=n}^{\infty} f^{(k)}(0),$$

and (1.7) defines the sequence $\{\mathscr{L}'_n\}$ given by

$$\mathscr{L}'_0(f) = 0,$$
$$\mathscr{L}'_n(f) = \sum_{k=0}^{n-1} f^{(k)}(0), \qquad n = 1, 2, \ldots.$$

The corresponding linear transformations $T = (\mathscr{L}_0, \mathscr{L}_1, \ldots)$ and $T' = (\mathscr{L}'_0, \mathscr{L}'_1, \ldots)$ are both right inverses for the transformation U associated with the sequence (1.5). The domain of T is a subspace of the domain of T', and so every function f in $\mathfrak{E}(\sigma, T)$ has two essentially different representations as series in the polynomials $\{p_n(z)\}$.

Further illustrations of these ideas will be given for other sets of polynomials as we come to them. Although results can be obtained by our methods for general basic sets of polynomials, the most interesting

---

[1] See also FALGAS [4], [5].

results apply only to basic sets that have a sufficient amount of intrinsic structure. We have found that a kind of generalized Appell set is sufficiently specialized to yield interesting results, yet sufficiently general to include many of the better-known polynomial sets, such as those associated with the names of LAGUERRE, LEGENDRE, HERMITE, CHE-BYSHEV, GEGENBAUER and JACOBI. Before introducing the class of polynomial sets with which we shall chiefly work, we turn to a discussion of integral representations for analytic functions.

§ 2. Representation formulas with a kernel

We shall use an integral representation for analytic functions which contains both the Cauchy integral formula and the Pólya representation for entire functions of exponential type. It will be developed here in a somewhat more general form than is required for the applications we shall make[1].

Let $K(z, w)$ be analytic for (z, w) in an open set Λ containing the plane $z = 0$. For any positive R, the compact set consisting of all points $(0, w)$ with $|w| \leq R$ lies in Λ, so that there is a positive δ such that (z, w) lies in Λ whenever $|z| \leq \delta$, $|w| \leq R$. For fixed R, let $\delta_1(R)$ be the supremum of such δ and set $\delta(R) = \lim_{h \to 0} \delta_1(R + h)$. Then, for any choice of R, all points (z, w) with $|z| < \delta(R)$ and $|w| < R$ lie in Λ.

Let $F(w)$ be regular for $|w| > R$ and let Γ be the circle $|w| = R + \varepsilon$. Then

$$f(z) = (2\pi i)^{-1} \int_{\Gamma} K(z, w) F(w) \, dw \qquad (2.1)$$

is regular for $|z| < \delta(R + 2\varepsilon)$. If we contract Γ by decreasing ε, we find that $f(z)$ is regular at least in the disk $|z| < \delta(R)$. Thus (2.1) defines a linear transformation T from the class of functions F that are regular at ∞ into the class of functions f that are regular at 0. We can also represent T as a sequence-to-sequence matrix transformation. Let

$$K(z, w) = \sum_{n, k=0}^{\infty} C_{n, k} z^n w^k, \qquad |z| < \delta(R), \ |w| < R,$$

and let

$$F(w) = \sum_{k=0}^{\infty} F_k w^{-k-1}, \qquad |w| > R.$$

[There is no loss of generality from assuming $F(\infty) = 0$, since $F(w) - F(\infty)$ yields the same $f(z)$ in (2.1) as $F(w)$ does.] Then

$$T(F)(z) = \sum_{n=0}^{\infty} z^n \sum_{k=0}^{\infty} C_{n, k} F_k = \sum_{n=0}^{\infty} a_n z^n = f(z). \qquad (2.2)$$

---

[1] For somewhat similar discussions see, in particular, A. J. MACINTYRE [1], [2], EVGRAFOV [1], [2], LOHIN [1], FALGAS [2].

If we write

$$\varphi = (F_0, F_1, \ldots), \qquad \alpha = (a_0, a_1, \ldots),$$

(2.2) is equivalent to the matrix equation $\alpha = C\varphi$, where $C = (C_{n,k})$.

Many familiar integral formulas have the form (2.1) or (2.2). The simplest example is obtained by taking C to be the identity matrix $I = (\delta_{n,k})$. Then

$$K(z, w) = \sum_{n,k=0}^{\infty} \delta_{n,k} z^n w^k = \sum_{n=0}^{\infty} (z w)^n = \frac{1}{1 - z w},$$

so that (2.1) becomes

$$f(z) = (2 \pi i)^{-1} \int_{\Gamma} \frac{F(w) \, dw}{1 - w z} \qquad (2.3)$$

In this case f has a simple alternative expression in terms of F, since $\alpha = I \varphi = \varphi$, $f(z) = z^{-1} F(z^{-1})$, and (2.3) can be written

$$F(z^{-1}) = (2 \pi i)^{-1} \int_{\Gamma} \frac{F(w) \, dw}{z^{-1} - w}$$

so that (2.1) reduces to CAUCHY's integral formula.

An interesting general class of transforms is obtained by restricting the matrix C to be triangular, with $C_{n,k} = 0$ for $n > k$. In this case $K(z, w)$ can be written

$$K(z, w) = \sum_{k=0}^{\infty} w^k \sum_{n=0}^{k} C_{n,k} z^n = \sum_{k=0}^{\infty} w^k Q_k(z),$$

where $Q_k(z)$ is a polynomial of degree k or less. Alternatively, we can write

$$K(z, w) = \sum_{n=0}^{\infty} z^n \sum_{k=n}^{\infty} C_{n,k} w^k$$

$$= \sum_{n=0}^{\infty} (z w)^n \sum_{k=0}^{\infty} C_{n, k+n} w^k$$

$$= \Psi(z w, w),$$

where $\Psi(s, t)$ is analytic in an open set containing the plane $s = 0$.

If $\Psi(s, t)$ is independent of t, then $K(z, w) = \Psi(z w)$, where $\Psi(t) = \sum \Psi_n t^n$ is regular at the origin. In this case, which is the one in which we are chiefly interested, the matrix C is diagonal, with $C_{n,n} = \Psi_n$. We then have

$$f(z) = (2 \pi i)^{-1} \int_{\Gamma} \Psi(z w) F(w) \, dw, \qquad (2.4)$$

or alternatively

$$\begin{aligned} \Psi(t) &= \sum_{n=0}^{\infty} \Psi_n t^n, \\ F(w) &= \sum_{n=0}^{\infty} F_n w^{-n-1}, \\ f(z) &= \sum_{n=0}^{\infty} F_n \Psi_n z^n. \end{aligned} \right\} \qquad (2.5)$$

If we assume further that no Ψ_n is zero, we can reconstruct $F(w)$ uniquely from a given $f(z)$, so that we can think of (2.4) as a representation formula for $f(z)$ instead of as defining a transform of $F(w)$. In this case it is convenient to change the notation and replace (2.5) by

$$\begin{aligned} \Psi(t) &= \sum_{n=0}^{\infty} \Psi_n t^n, \qquad \Psi_n \neq 0; \\ f(z) &= \sum_{n=0}^{\infty} f_n z^n, \\ F(w) &= \sum_{n=0}^{\infty} \frac{f_n}{\Psi_n w^{n+1}} . \end{aligned} \right\} \qquad (2.6)$$

When $\Psi(t) = e^t$, (2.6) describes the correspondence (PÓLYA [1]) between an entire function $f(z)$ of exponential type and its Laplace (or Borel) transform $F(w)$. The general case can be applied, by suitable choice of Ψ, either to entire functions of arbitrary order or to functions that are regular in a prescribed region.

The representation provided by (2.4), (2.6) is most convenient to use when $\Psi(t)$ is restricted by auxiliary conditions on its coefficients Ψ_n. We call $\Psi(t)$ a *comparison function* if $\Psi_n > 0$ and $\Psi_{n+1}/\Psi_n \downarrow 0$. A comparison function is necessarily entire, as the ratio test for convergence shows. When $\Psi(t)$ is a comparison function, we denote by $\Re_\Psi$ the class of entire functions f such that, for some number τ (depending on f),

$$|f(r e^{i\theta})| \leq M \Psi(\tau r), \qquad r \uparrow \infty . \qquad (2.7)$$

We call $\Re_\Psi$ the class of functions of finite Ψ-type. The infimum of numbers τ for which (2.7) holds is the (exact) Ψ-type of f; we denote by $\Re_\Psi(\tau)$ the class of functions whose Ψ-type is τ or less. For example, when $\Psi(t) = e^t$, $\Re_\Psi(\tau)$ is the class of functions of exponential type τ, that is, entire functions of order 1 and type not exceeding τ, or of order less than 1.

The Ψ-type of a function can be computed from the coefficients in its power series by applying the following theorem (NACHBIN [1]).

NACHBIN's theorem. *A function* $f(z) = \sum_{n=0}^{\infty} f_n z^n$ *is of* Ψ-type τ *if and only if* $\limsup |f_n/\Psi_n|^{1/n} = \tau$.

For the convenience of the reader, we give a proof here. First, let $\lim \sup |f_n/\Psi_n|^{1/n} = \tau < \infty$. Then, if $\tau_1 > \tau$, we may choose B so that $|f_n/\Psi_n| \leq B \tau_1^n$ for $n = 0, 1, 2, \ldots$. Thus, on the circle $|z| = r$,

$$|f(z)| \leq \sum_{n=0}^{\infty} |f_n| r^n \leq B \sum_{n=0}^{\infty} \tau_1^n \Psi_n r^n = B \Psi(\tau_1 r).$$

Since τ_1 may be arbitrarily close to τ, this shows that f is of Ψ-type at most τ.

In the other direction, we need a simple lemma connecting the rate of growth of Ψ with that of its coefficients.

Lemma. *Let* $\gamma_n = \min_{x>0} \Psi(x) x^{-n}$. *Then, for all nonnegative integers* n,

$$1 \leq \gamma_n/\Psi_n \leq (n+1) e, \tag{2.8}$$

and consequently $\lim (\gamma_n/\Psi_n)^{1/n} = 1$.

Since $\Psi(x) \geq \Psi_n x^n$, it is evident that $\gamma_n \geq \Psi_n$. To obtain the right-hand side of (2.8), we estimate $\Psi(x)$ for a choice of x near that which minimizes $\Psi(x) x^{-n}$. Let $d_n = \Psi_{n-1}/\Psi_n$ and let $0 < \omega < 1$; ω is to be near 1, and will be specified later. Recalling that a restriction on Ψ was that $\{d_n\}$ increases, we observe that $\Psi_k \leq \Psi_n d_n^{n-k}$, both for $k < n$ and $k \geq n$. Setting $x = \omega d_n$, we have

$$\Psi(x) = \sum \Psi_k x^k \leq \Psi_n \sum_0^{\infty} d_n^{n-k} (\omega d_n)^k$$

$$\leq \Psi_n d_n^n/(1 - \omega).$$

For this choice of x, we have $\Psi(x)/x^n \leq \Psi_n \omega^{-n}/(1 - \omega)$, and so $\gamma_n/\Psi_n \leq \omega^{-n}/(1 - \omega)$. Choosing ω as $n/(n+1)$ to minimize the right-hand side, we obtain

$$\gamma_n/\Psi_n \leq (n+1)(1 + n^{-1})^n \leq (n+1) e.$$

To apply the lemma, suppose that f is of Ψ-type τ. If $\tau_1 > \tau$, then for some M we have $|f(z)| \leq M \Psi(\tau_1 r)$. By CAUCHY's inequality,

$$|f_n| \leq M \Psi(\tau_1 r) r^{-n} = M \tau_1^n \Psi(\tau_1 r)(\tau_1 r)^{-n}.$$

Choosing r to minimize the right-hand side, we find $|f_n| \leq M \tau_1^n \gamma_n$, and

$$|f_n/\Psi_n|^{1/n} \leq M^{1/n} \tau_1 (\gamma_n/\Psi_n)^{1/n}.$$

Invoking the lemma, we then have $\lim \sup |f_n/\Psi_n|^{1/n} \leq \tau_1$. The conclusion of NACHBIN's theorem follows on letting τ_1 approach τ.

Now let $f \in \mathfrak{R}_\Psi$, let F be defined by (2.6), and let $D(f)$ denote the union of the set of all singular points of F and the set of all points exterior to the domain of F. The contour Γ in (2.4) can then be any contour enclosing $D(f)$. If $f \in \mathfrak{R}_\Psi(\tau)$, then $D(f)$ lies in the disk $|w| \leq \tau$ and Γ may be taken as the circle $|w| = \varrho > \tau$. Additional information about $D(f)$ will lead to more detailed estimates of the growth of f in various directions.

We summarize the relevant parts of the preceding discussion in a formal theorem.

Theorem 2.9. *Let* $\Psi(t) = \sum\limits_0^\infty \Psi_n t^n$ *be a comparison function, i.e.* $\Psi_n > 0$ *and* $\Psi_{n+1}/\Psi_n \downarrow 0$. *Let* $f(z) = \sum\limits_{n=0}^\infty f_n z^n$ *belong to the class* $\Re_\Psi$ *[as in (2.7)], and let* $D(f)$ *be the closed set described in the preceding paragraph. Then*

$$f(z) = \frac{1}{2\pi i} \int_\Gamma \Psi(zw) F(w) \, dw$$

where Γ *encloses* $D(f)$ *and*

$$F(w) = \sum_{n=0}^\infty \frac{f_n}{\Psi_n w^{n+1}}. \qquad (2.10)$$

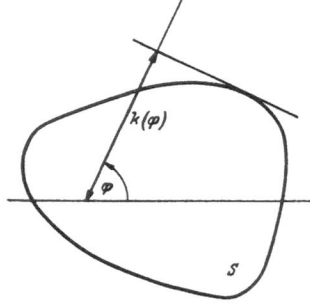

Fig. 1. The supporting function for a convex set

When $\Psi(t)$ is chosen as e^t, this theorem becomes the familiar representation for entire functions of exponential type. If $f(z) = \sum\limits_{n=0}^\infty a_n z^n/n!$ is entire, and of growth at most order 1, finite type, then

$$f(z) = \frac{1}{2\pi i} \int_\Gamma e^{zw} F(w) \, dw,$$

where $F(w) = \sum\limits_{n=0}^\infty a_n/w^{n+1}$ and Γ encircles the set $D(f)$. If f is of exponential type τ, then $D(f)$ lies in the closed disk $|w| \leq \tau$. PÓLYA, in his notable memoir [1], showed that the rate of growth of f along radial lines sharply delimits the set $D(f)$. (A detailed treatment is to be found in BOAS [3], BIEBERBACH [2], CARTWRIGHT [1] or LEVIN [1].)

With any closed set S in the plane, one may associate a supporting function

$$k(\varphi) = k(\varphi; S) = \sup_{z \in S} \Re(z e^{-i\varphi}).$$

If S^Δ is the closed convex hull of S, then $k(\varphi; S) = k(\varphi; S^\Delta)$. (See Fig. 1.)

With an entire function f of order 1, one may associate an indicator (growth) function

$$h(\theta) = h(\theta; f) = \limsup r^{-1} \log |f(r e^{i\theta})|.$$

If f is of finite type τ, then $h(\theta; f) \leq \tau$ for every θ.

The central fact discovered by PÓLYA was that $h(\theta)$ is also the supporting function for a convex set, namely, the conjugate of the set

$D(f)^\Delta$. This is called the conjugate indicator diagram of f. This relation can be stated concisely as follows: if f is any entire function of exponential type, then $k\left(\theta; D(f)^\Delta\right) = h(-\theta; f)$ for each θ. The proof depends upon the fact that (2.10) can be given an integral form

$$F(w) = w^{-1} \int_0^\infty e^{-t} f(t/w)\, dt$$
$$= \int_0^\infty e^{-ws} f(s)\, ds$$

from which it easily follows that $F(w)$ is analytic in each half plane $\Re(w\, e^{-i\theta}) > h(-\theta; f)$.

As an illustration of the way in which the relation between $h(\theta)$ and $D(f)$ is used, observe that if f obeys $h(\pm \pi/2) \leq c$, then $D(f)$ lies in the strip $|v| \leq c$.

For a general comparison function $\Psi(t)$, the relationship between $D(f)$ and growth rates of f is not so precise. Some information can be achieved about the shape of $D(f)$. It is again possible to obtain an integral form for the generalized Borel transform (2.10). Choose a function $\alpha(t)$, of bounded variation on the interval $[0, \infty)$, so that

$$1/\Psi_n = \int_0^\infty t^n\, d\alpha(t). \tag{2.11}$$

Then,

$$F(w) = w^{-1} \int_0^\infty f(t/w)\, d\alpha(t). \tag{2.12}$$

[In § 13, we shall use an analogous approach in dealing with the case in which $\Psi(t)$ is not entire.]

In this direction, the sharpest results have been obtained by A. J. MACINTYRE [1], using again a specialized choice of $\Psi(t)$. If we wish to discuss entire functions of order ϱ, it would be appropriate to choose $\Psi(t)$ as a function of order ϱ, type 1, whose coefficients Ψ_n have a simple form. Two natural choices are $\sum t^n/(n!)^{1/\varrho}$ and $\sum t^n/\Gamma(1 + n/\varrho)$, which reduce to e^t for $\varrho = 1$. Instead of these, MACINTYRE chose $\Psi(t) = \sum t^n/\Gamma\left((1+n)\varrho^{-1}\right)$. For this, (2.12) takes the form

$$1/\Psi_n = \Gamma(\varrho^{-1} + n\varrho^{-1}) = \int_0^\infty t^n\, \varrho\, e^{-t^\varrho}\, dt$$

so that

$$F(w) = \frac{\varrho}{w} \int_0^\infty f(t/w)\, e^{-t^\varrho}\, dt. \tag{2.13}$$

Introduce the analogous growth function

$$h_\varrho(\theta) = \limsup r^{-\varrho} \log|f(r\, e^{i\theta})|.$$

(In general, this is not the supporting function for a convex set.) It is then immediate that $F(w)$ is analytic at $w = R e^{i\varphi}$ where $R > |h_\varrho(-\varphi)|^{1/\varrho}$. This confines the set $D(f)$, outside of which F is regular, to a certain star-shaped set about the origin. With a change of variable, (2.13) becomes

$$F(w^{1/\varrho} e^{i\theta}) = \varrho e^{-i\theta} \int_0^\infty f(s\, e^{-i\theta})\, e^{-w s^\varrho}\, ds,$$

from which it follows that $F(w^{1/\varrho} e^{i\theta})$ is analytic when w lies in the half plane $\Re(w) > h_\varrho(-\theta)$, for each θ. This sharpens the previous estimate of $D(f)$, taking advantage of possible negative values of $h_\varrho(-\theta)$. [One could hope for a simple geometrical characterization of the set $D(f)$, or some set obtained from it, similar to that obtained by PÓLYA for $\varrho = 1$.]

In the discussion to follow, we shall often make use of the indicator function and conjugate indicator diagram relation for functions of exponential type, but not for functions of order different from 1, since in the examples which make up the bulk of Chapter II we encounter chiefly the former.

§ 3. The method of kernel expansion

Suppose that we want to represent a given function $f(z)$ in the form of a series $\sum c_n p_n(z)$, with a prescribed sequence of functions $p_n(z)$, which need not, for the present, be supposed to be polynomials. Let us choose a suitable sequence of functions $u_n(w)$ and form the function

$$K(z, w) = \sum p_n(z)\, u_n(w) \qquad (3.1)$$

assuming that this series converges for (z, w) in an open set Λ containing the plane $z = 0$. We can then use $K(z, w)$ as the kernel of a transformation $\boldsymbol{T}$ in accordance with (2.1). Let us further assume that (3.1) converges uniformly in w for w on a simple closed contour Γ, and some range of z; we then refer to Γ as an admissible path. If $f = \boldsymbol{T}(F)$, with F regular on the admissible path Γ, we have

$$f(z) = \frac{1}{2\pi i} \int_\Gamma K(z, w)\, F(w)\, dw$$

$$= \frac{1}{2\pi i} \int_\Gamma \sum p_n(z)\, u_n(w)\, F(w)\, dw$$

$$= \sum p_n(z) \frac{1}{2\pi i} \int_\Gamma u_n(w)\, F(w)\, dw$$

for each z for which Γ is admissible. Thus

$$f(z) = \sum p_n(z)\, \mathscr{L}_n(f), \qquad (3.2)$$

where
$$\mathscr{L}_n(f) = \frac{1}{2\pi i} \int_\Gamma u_n(w)\, F(w)\, dw, \tag{3.3}$$

and we have an expansion of the desired form, with an explicit formula for the coefficients.

The same equations (3.1), (3.2), (3.3) can also be reached by a quite different route. Suppose we are presented with a sequence of functionals $\{\mathscr{L}_n\}$ applicable to a certain class of analytic functions. We pose the following interpolation problem: express f in terms of the sequence of numbers $\{\mathscr{L}_n(f)\}$. Let us suppose that for this class of analytic functions, we have an appropriate representation formula

$$f(z) = \frac{1}{2\pi i} \int_\Gamma K(z, w)\, F(w)\, dw$$

with a known kernel $K(z, w)$. The linear functionals $\{\mathscr{L}_n\}$ will then have integral representations of the form (3.3) in terms of an associated sequence of functions $\{u_n(w)\}$. With these known, we next seek functions $p_n(z)$ so that (3.1) holds. Repeating the process outlined in the preceding paragraph, we again arrive at (3.2), which now expresses the function f in terms of the preassigned numbers $\{\mathscr{L}_n(f)\}$.

The effective use of either of these procedures for a given class of functions f depends upon our ability to select an appropriate kernel $K(z, w)$; nor is the kernel in any way unique, although one may be more effective than another. We illustrate this with a trivial example: the sequence $\{z^n\}$ is the sequence $\{p_n(z)\}$ associated in the sense of (3.1) with (among others)

$$K(z, w) = \sum_{n=0}^{\infty} z^n w^n = \frac{1}{1 - z w} \tag{3.4}$$

or with

$$K(z, w) = \sum_{n=0}^{\infty} z^n w^n / n! = e^{z w}. \tag{3.5}$$

Using (3.4), $F(w) = w^{-1} f(1/w)$, and an admissible path is one on which $|z w| < 1$. By the argument outlined above, it follows that any $f(z)$ regular at 0 has an expansion of the form $\sum c_n z^n$, converging uniformly in compact subsets of the largest open disk (center at 0) in which f is regular. This is, of course, just TAYLOR's theorem. On the other hand, if we use the kernel of (3.5), any closed contour is admissible; however, since the kernel $e^{z w}$ generates the Pólya representation, the only functions f that are obtained in the form $T(F)$ are entire functions of exponential type, and the process furnishes only the much weaker conclusion that such functions are representable by power series $\sum c_n z^n / n!$.

The example (3.4) also illustrates a less obvious point. While (3.4) converges only when $|z w| < 1$, it is summable in much larger domains by appropriate methods, such as Borel or Mittag-Leffler summability[1]. A general class of useful summability methods can be obtained as follows. Let $E(t) = \sum C_n t^n$ be an entire function, with $C_n > 0$ for $n = 0, 1, \ldots$. If $\{s_n\}$ is a complex sequence that does not grow too fast $\left(|s_n|^{1/n} = O(1)\right)$ we define

$$\text{E-Lim } s_n = \lim_{t \uparrow \infty} \frac{\sum s_n C_n t^n}{E(t)}$$

For a series, (E)-$\sum A_n = S$ means E-Lim $s_n = S$, where $\{s_n\}$ is the sequence of partial sums of $\sum A_n$. Let $\mathscr{E}$ be the set of points z such that $\lim_{t \uparrow \infty} E(zt)/E(t) = 0$. This is a star set containing the open unit disk, but not the point 1. It is easily seen that $\mathscr{E}$ is precisely the set of z for which (E)-$\sum_{0}^{\infty} z^n = 1/(1-z)$. For example, with $E(t) = e^t$ (Borel summability), $\mathscr{E}$ is the half plane $\Re(z) < 1$.

Returning to the series (3.4), we have

$$\text{(E)-}\sum z^n w^n = 1/(1-zw)$$

for all (z, w) with $zw \in \mathscr{E}$, with uniform summability when zw is confined to a compact subset of $\mathscr{E}$. Suppose that f is any function regular in a simply connected region Ω containing the origin. Form the set[2] $\Omega_0 = (\Omega' \cdot \mathscr{E}')'$. Then Ω_0 is an open star-shaped subset of Ω, called by Bieberbach [2] the product star of Ω and $\mathscr{E}$, and denoted by $\Omega \odot \mathscr{E}$. If G is a compact subset of Ω_0, we may choose compact sets $\Omega_1 < \Omega$ and $\mathscr{E}_1 < \mathscr{E}$ so that $G < \Omega_1 \odot \mathscr{E}_1$. Let Γ_1 be a closed contour in Ω surrounding Ω_1, and let Γ be $1/\Gamma_1$. Then, for any $z \in \Omega_1$,

$$f(z) = \frac{1}{2\pi i} \int_{\Gamma} \frac{w^{-1} f(1/w)}{1 - zw} \, dw.$$

If z is in G and w on Γ (so that $1/w$ is in Ω_1') then $z = (zw)(1/w)$ lies in the complement of $\Omega_1' \cdot \mathscr{E}_1'$, and zw lies in $\mathscr{E}_1$. Hence, Γ is an admissible path for E-summability, that is, (E)-$\sum z^n w^n = 1/(1-zw)$ uniformly for all $z \in G$ and w on Γ. Proceeding as before, we substitute and integrate, getting

$$f(z) = \text{(E)-}\sum z^n \frac{1}{2\pi i} \int_{\Gamma} w^{n-1} f(1/w) \, dw$$

$$= \text{(E)-}\sum z^n f^{(n)}(0)/n!.$$

What we have proved, of course, is the familiar fact that the power series for $f(z)$ is E-summable to $f(z)$ uniformly in every compact subset

---

[1] Dienes [1], p. 311.

[2] We use S' to mean the complement of S, $1/S$ to mean the set of points $1/z$ with $z \in S$, and $S_1 \cdot S_2$ to mean the set of all points of the form $z_1 z_2$ with $z_1 \in S_1$, $z_2 \in S_2$.

of the set $\Omega \odot \mathscr{E}$, the E-star of the region of regularity of f. When $E(t) = e^t$, $\Omega \odot \mathscr{E}$ is the Borel polygon (star) of Ω, obtained as the intersection of half planes whose bounding lines are orthogonal to the rays from the origin to boundary points of Ω. When $E(t)$ is $\sum t^n/\Gamma(1 + \sigma n)$, $\mathscr{E}$ is the set of $z = 1 + r e^{i\theta}$ with $r > 0$ and $\pi \geq |\theta| > \sigma \pi > 0$. If $E(t)$ is chosen as either $\sum t^n/\Gamma(1 + n [\log n]^{-\frac{1}{2}})$ or as $\sum t^n/[\log (n + 2)]^n$, then $\mathscr{E}$ is the whole plane, except for the portion of the real axis $x \geq 1$. In this case, the product star $\Omega \odot \mathscr{E}$ is the Mittag-Leffler star of Ω, obtained from Ω by deleting the rays extending from boundary points of Ω to infinity. (See DIENES [1], p. 311 or COOKE [1], p. 182.) In either of the last two cases we refer to E-summability as Mittag-Leffler summability.

The same process can be applied in the general case. If the kernel expansion (3.1) is E-summable uniformly for w on a path Γ, we say that Γ is admissible for E-summability. By the same reasoning, we obtain (3.2) again, as an E-summable representation of $f(z)$. Depending upon the point of view, this is either a summable expansion of $f(z)$ in terms of the preassigned functions $\{p_n(z)\}$, or a summable representation of $f(z)$ in terms of the preassigned functional values $\{\mathscr{L}_n(f)\}$.

§ 4. Lidstone series [1]

In this and the next section, we illustrate the technique explained in the preceding section with two simple examples. Let us consider the problem of representing an entire function of exponential type in terms of the values of its even derivatives at the points 0 and 1. This is clearly an interpolation problem. From the Pólya representation

$$f(z) = \frac{1}{2\pi i} \int_\Gamma e^{zw} F(w)\, dw$$

we have

$$f^{(2n)}(0) = \frac{1}{2\pi i} \int_\Gamma w^{2n} F(w)\, dw,$$

$$f^{(2n)}(1) = \frac{1}{2\pi i} \int_\Gamma e^w w^{2n} F(w)\, dw,$$

where Γ is a contour surrounding the conjugate indicator diagram $D(f)$ of f. The kernel $K(z, w)$ is e^{zw}, so that we seek an expansion of the form (3.1), which in this case is

$$e^{zw} = \sum w^{2n} p_n(z) + \sum e^w w^{2n} q_n(z).$$

We may start from the observation that this would require that $e^{zw} = A(w^2, z) + e^w B(w^2, z)$. Replacing w by $-w$, and solving the

---

[1] BOAS [1], BUCK [5]; further references are given in these papers.

resulting pair of equations, we obtain the identity

$$e^{zw} = \frac{\sinh (1-z)\,w}{\sinh w} + e^w\,\frac{\sinh z\,w}{\sinh w}.$$

We may write

$$\frac{\sinh z\,w}{\sinh w} = \sum_{n=0}^{\infty} A_n(z)\,w^{2n}, \tag{4.1}$$

where the coefficient $A_n(z)$ is a polynomial. Since the left side is regular except for simple poles at $w = k\pi i$, $k = \pm 1, \pm 2, \ldots$, this series converges uniformly in any compact subset of the disk $|w| < \pi$. This gives the desired expansion of the kernel, valid for $|w| < \pi$:

$$e^{zw} = \sum_{n=0}^{\infty} w^{2n} A_n(1-z) + \sum_{n=0}^{\infty} e^w\,w^{2n} A_n(z). \tag{4.2}$$

Any circle $|w| = \varrho < \pi$ is an admissible path Γ; the corresponding functions f must be such that $D(f)$ lies inside Γ. Using the Pólya correspondence, this means that we can apply the procedure of §3 to functions f of exponential type τ for any $\tau < \varrho < \pi$, and we have proved the following result.

Theorem 4.3. *Any entire function $f(z)$ of exponential type less than π has a convergent Lidstone representation*

$$f(z) = \sum_{n=0}^{\infty} A_n(1-z)\,f^{(2n)}(0) + \sum_{n=0}^{\infty} A_n(z)\,f^{(2n)}(1).$$

Returning to (4.1), we notice that this series is Borel summable in the strip $|v| < \pi$. The procedure may therefore be applied to any function f whose indicator set $D(f)$ lies in this strip; employing the connection between $h(\theta; f)$ and $D(f)$, we immediately obtain

Theorem 4.4. *Any entire function $f(z)$ of exponential type which obeys $h(\pm \pi/2; f) < \pi$ has a Borel-summable Lidstone series.*

Again, (4.1) is Mittag-Leffler summable for all values of $w = u + iv$ except those with $u = 0$, $|v| \geq \pi$. If $D(f)$ avoids these points, then the Lidstone series for $f(z)$ is ML-summable to $f(z)$. This condition is certainly satisfied if $D(f)$ lies in either the open left half plane, or in the open right half plane.

Theorem 4.5. *Any entire function $f(z)$ of exponential type which obeys $h(0; f) < 0$, or $h(\pi; f) < 0$, has a Mittag-Leffler summable Lidstone series.*

Suppose now that f is merely a function of exponential type $\tau < \infty$. The set $D(f)$ is then a subset of the disk $|w| \leq \tau$. Consider the function $g(z) = e^{cz} f(z)$; it is easily verified that $D(g)$ is the set $D(f) + c$, and therefore lies in the disk with center at c, and radius τ. If c is large and positive, $D(g)$ lies in the right half plane. Set

$$\mathscr{L}_n(f) = g^{(2n)}(0) = \sum_{k=0}^{2n} \binom{2n}{k} c^{2n-k} f^{(k)}(0),$$

$$\mathscr{M}_n(f) = g^{(2n)}(1) = e^c \sum_{k=0}^{2n} \binom{2n}{k} c^{2n-k} f^{(k)}(1).$$

We may then apply Theorem 4.5 to g, and obtain

Corollary. *If f is any function of exponential type τ, and $c > \tau$, then*

$$f(z) = e^{-cz}\{(\mathrm{ML})\text{-}\sum A_n(1-z)\,\mathscr{L}_n(f) + (\mathrm{ML})\text{-}\sum A_n(z)\,\mathscr{M}_n(f)\}.$$

[Although the polynomials $A_n(z)$ are those of Theorem 4.3, and are defined by (4.1), this is not a Lidstone series.]

The type restriction in Theorem 4.3 is best possible, as may be seen from $f(z) = \sin \pi z$. However, (4.1) can be modified to allow an extension to functions of arbitrary type. The convergence of (4.1) stops at the simple poles $\pm \pi i$. Evaluating the residues there, we have

$$\frac{\sinh z w}{\sinh w} = 2\pi \frac{\sin \pi z}{w^2 + \pi^2} + G(z, w),$$

where $G(z, w)$ is regular for $|w| < 2\pi$, and is even in w. More generally,

$$\frac{\sinh z w}{\sinh w} = 2 \sum_{k=1}^{N} \frac{(-1)^{k+1} k \pi \sin(k \pi z)}{w^2 + k^2 \pi^2} + G(z, w),$$

where $G(z, w) = \sum B_n(z) w^{2n}$, convergent for $|w| < (N+1)\pi$. The function $B_n(z)$ is no longer a polynomial, but is the sum of $A_n(z)$ and a certain trigonometric polynomial. This in turn gives

$$e^{zw} = \sum_{n=0}^{\infty} B_n(1-z) w^{2n} + \sum_{n=0}^{\infty} e^w w^{2n} B_n(z) + H(z, w),$$

where the last term is a finite sum

$$H(z, w) = \sum_{k=1}^{N} k \pi \sin(k \pi z) \frac{1 + (-1)^{k+1} e^w}{w^2 + k^2 \pi^2},$$

and the kernel expansion converges for $|w| < (N+1)\pi$, and all z. The routine application of the basic procedure yields finally (BUCK [5]).

Theorem 4.6. *If f is any entire function of finite type τ, then*

$$f(z) = \sum_{n=0}^{\infty} B_n(1-z) f^{(2n)}(0) + \sum_{n=0}^{\infty} B_n(z) f^{(2n)}(1)$$

$$+ \sum_{k=1}^{N} C_k \sin(k \pi z),$$

where $N \pi \leq \tau$.

Since the focus of attention has been upon the functionals, it is not disconcerting that this expansion uses functions $B_n(z)$ which differ from the original polynomials $A_n(z)$. For example, we may use this last result to obtain a result of I. J. SCHOENBERG's [1]: *if f is an entire function of finite exponential type, and $f^{(2n)}(0) = f^{(2n)}(1) = 0$ for $n = 0$, 1, 2, ..., then $f(z)$ is a finite trigonometric sum $\sum C_k \sin(k \pi z)$.*

§ 5. A set of Laguerre polynomials

We shall now take up a simple case of the expansion problem, using as an illustration a simple family of polynomial sets which includes that used as an illustration in § 1, and shows how the multiple expansions observed there fit into the theory of § 3. We take

$$(1 - w)^\lambda e^{zw} = \sum_{n=0}^{\infty} w^n p_n(z),$$

with the principal value of the power when λ is not integral; then[1]

$$p_n(z) = (-1)^n L_n^{(\lambda - n)}(z),$$

where $L_n^{(m)}$ is the Laguerre polynomial of degree n and index m. Here we consider only the case when λ is a positive integer [cf. § 9 (ii) and § 18]. Taking $K(z, w) = e^{zw}$, we have

$$e^{zw} = \sum_{n=0}^{\infty} (1 - w)^{-\lambda} w^n p_n(z).$$

Any path Γ not passing through $w = 1$ is admissible. If $f(z)$ is an entire function of exponential type τ, and Γ surrounds the circle $|z| = \tau$, we have, by (3.2) and (3.3),

$$f(z) = \sum_{n=0}^{\infty} p_n(z) \mathscr{L}_n(f), \tag{5.1}$$

with

$$\mathscr{L}_n(f) = (2 \pi i)^{-1} \int_\Gamma (1 - w)^{-\lambda} w^n F(w) \, dw. \tag{5.2}$$

If $\tau > 1$, the singular points of the integrand in (5.2) are all inside Γ and $\mathscr{L}_n(f)$ is independent of how Γ is chosen. However, if $\tau < 1$, a Γ that does not have $w = 1$ inside it, and a Γ that does, give rise to different expansion formulas because of the singular point at $w = 1$. In particular, when $\lambda = 1$ the generating relation becomes

$$(1 - w) e^{zw} = \sum_{n=0}^{\infty} p_n(z) w^n,$$

---

[1] ERDÉLYI [1], vol. 2, p. 189, (19).

so that the $p_n(z)$ are the negatives of the polynomials (1.5). We then have

$$\mathscr{L}_n(f) = \frac{1}{2\pi i} \int_\Gamma \frac{w^n F(w)}{1-w}\, dw;\qquad (5.3)$$

if $w=1$ is inside Γ, (5.3) is

$$\mathscr{L}_n^{(1)}(f) = -\sum_{j=0}^{n-1} f^{(j)}(0),\qquad (5.4)$$

while if $w=1$ is outside Γ, (5.3) is

$$\mathscr{L}_n^{(2)}(f) = \sum_{j=n}^{\infty} f^{(j)}(0).\qquad (5.5)$$

If $f(z)$ is an entire function of exponential type less than 1, we have the two convergent expansions

$$\left.\begin{aligned}
f(z) &= \sum_{n=0}^{\infty} p_n(z)\,\mathscr{L}_n^{(1)}(f),\\
f(z) &= \sum_{n=0}^{\infty} p_n(z)\,\mathscr{L}_n^{(2)}(f),
\end{aligned}\right\}\qquad (5.6)$$

and these are different unless $\sum_{j=0}^{\infty} f^{(j)}(0) = 0$, i.e. $F(1) = \int_0^\infty e^{-x} f(x)\, dx = 0$. Except in the latter case we obtain $\sum p_n(z) \equiv 0$ by subtracting the two expansions (5.6), a relation that we also can obtain by taking $w=1$ in the generating relation for the $p_n(z)$.

Let Φ be any linear functional on the class of entire functions of exponential type. Then we have

$$\sum_{n=0}^{\infty} \Phi(f)\, p_n(z) = 0,$$

and so

$$f(z) = \sum_{n=0}^{\infty} p_n(z)\, \{\mathscr{L}_n^{(1)}(f) + \Phi(f)\}.$$

Thus there are infinitely many distinct expansions of any f [except when $f(z) \equiv 0$]. In § 8 we shall see that there cannot be more than two independent expansions $0 = \sum c_n p_n(z)$ with $c_n = O(R^n)$ for some R.

We thus have a complete explanation of the facts about (1.5) observed in § 1, as far as they concern entire functions of exponential type.

§ 6. Generalized Appell polynomials

We now introduce a class of polynomials which, while sufficiently general to include many standard sets, is still sufficiently specialized to have a simple theory, and is well adapted to the application of our general method.

Definition. *A set of polynomials has a generalized Appell representation if it is generated by the formal relation*

$$A(w)\,\Psi\big(z\,g(w)\big) = \sum_{n=0}^{\infty} p_n(z)\,w^n, \tag{6.1}$$

where

$$\left.\begin{array}{ll} A(w) = \displaystyle\sum_{n=0}^{\infty} a_n w^n, & a_0 \neq 0; \\[2mm] \Psi(t) = \displaystyle\sum_{n=0}^{\infty} \Psi_n t^n, & no\ \Psi_n = 0; \\[2mm] g(w) = \displaystyle\sum_{n=1}^{\infty} g_n w^n, & g_1 \neq 0. \end{array}\right\} \tag{6.2}$$

If a set of polynomials has a generalized Appell representation we shall refer to the polynomials as generalized Appell polynomials. Sets of polynomials generated by the more general kernel $A(w)\,\Psi\{(a+bz)\,g(w)\}$ occur occasionally; their theory is reduced to that considered here by a linear transformation (cf. § 17).

It is sometimes possible to change a given polynomial set $\{p_n\}$ into one of generalized Appell type by multiplying each p_n by a suitable non-zero constant c_n; of course, the expansion properties of $\{p_n\}$ and $\{c_n p_n\}$ are the same. However, we do not know how to predict when such a sequence of multipliers exists. Some sets of polynomials have more than one representation of generalized Appell type [cf. (9.8) and (10.8)]; again, we do not know either necessary or sufficient conditions for this to happen.

It is easy to see that $p_n(z)$ is in fact a polynomial of degree n [cf. (6.4), below]. The choice $g(w) = w$, $\Psi(t) = e^t$ gives Appell polynomials; $g(w) = w$ gives a class that has been studied by BRENKE [1], HUFF [1], and HUFF and RAINVILLE [1] and that we shall call the class of Brenke polynomials; $\Psi(t) = e^t$ gives what we shall call Sheffer polynomials (see § 10)[1].

Appell polynomials can be characterized by a well-known explicit representation in terms of the coefficients a_n, as well as by the recursion relation $p_n'(z) = p_{n-1}(z)$. We shall characterize generalized Appell polynomials by an explicit formula and a recursion relation (BOAS and BUCK [1]).

Theorem 6.3. *The polynomials generated by* (6.1) *have the explicit representation*

$$p_n(z) = \sum_{j=0}^{n} z^j \Psi_j \sum a_{k_0} g_{k_1} g_{k_2} \cdots g_{k_j}, \tag{6.4}$$

---

[1] The polynomials obtained when z is replaced by a polynomial $f(z)$ in (6.1) and $\Psi(t) = e^t$ were studied by PALAS [1].

*where the inner summation extends over all sets of $j+1$ nonnegative integers
$\{k\}$ such that $k_0+k_1+\cdots+k_j=n$; the coefficients are those appearing
in (6.2).*

When $g(w)=w$, (6.4) reduces to a formula of BRENKE's [1], and if
in addition $\Psi(t)=e^t$, it becomes the well-known formula

$$p_n(z) = \sum_{j=0}^{n} z^j a_{n-j}/j!$$

for Appell polynomials[1].

To prove Theorem 6.3, write $K(z,w) = A(w)\,\Psi\big(zg(w)\big)$. Let $p_n(z) = \sum_{j=0}^{n} p_{nj} z^j$. Then

$$K(z,w) = \sum_{n=0}^{\infty} w^n \sum_{j=0}^{n} p_{nj} z^j = \sum_{j=0}^{\infty} z^j \sum_{n=j}^{\infty} p_{nj} w^n,$$

so that

$$k! \sum_{n=j}^{\infty} p_{nj} w^n = \partial^j K/\partial z^j \Big|_{z=0}.$$

We also have

$$\partial^j K/\partial z^j = A(w)\,\{g(w)\}^j\,\Psi^{(j)}\big(z\,g(w)\big),$$

and so

$$\sum_{n=j}^{\infty} p_{nj} w^n = \Psi_j\,A(w)\,\{g(w)\}^j.$$

Picking out the coefficient of w^n in the product of the $j+1$ power
series on the right, we obtain the coefficient of z^j in (6.4).

If $\{p_n(z)\}$ is now any sequence of polynomials with p_n of degree n,
form the series

$$\sum_{n=0}^{\infty} w^n p_n(z) = K(z,w).$$

We shall characterize, in terms of properties of K, sets $\{p_n\}$ of generalized
Appell polynomials.

Theorem 6.5. *A necessary and sufficient condition that $K(z,w) = A(w)\,\Psi\big(zg(w)\big)$ with A, Ψ, g as in (6.2) and $g_1=1$ is that there exist power
series*

$$\left. \begin{array}{l} c(w) = \displaystyle\sum_{n=0}^{\infty} c_n w^n, \\[2mm] b(w) = 1 + \displaystyle\sum_{n=1}^{\infty} b_n w^n \end{array} \right\} \tag{6.6}$$

such that

$$K_2(z,w) = c(w)\,K(z,w) + z\,w^{-1}\,b(w)\,K_1(z,w), \tag{6.7}$$

where subscripts denote partial derivatives.

If $K(z,w)$ has the specified form, we can verify by differentiation
that (6.7) holds with $c(w) = A'(w)/A(w)$, $b(w) = w\,g'(w)/g(w)$.

---

[1] For Sheffer polynomials, another representation is given by SHEFFER [5].

Conversely, if (6.7) holds, choose $A(w)$ as in (6.2) so that $A'(w)/A(w) = c(w)$. Set $H(z, w) = K(z, w)/A(w)$. Since $K_1(z, w) = A(w) H_1(z, w)$ and

$$K_2(z, w) = A'(w) H(z, w) + A(w) H_2(z, w),$$

(6.7) implies that

$$H_2(z, w) = z w^{-1} b(w) H_1(z, w). \qquad (6.8)$$

Now choose $g(w)$ as in (6.2) so that $b(w) = w g'(w)/g(w)$, and so that $g(0) = 0$, $g'(0) = 1$. Then (6.8) states that

$$H_2(z, w) = \{z g'(w)/g(w)\} H_1(z, w).$$

Replacing z by $z/g(w)$, we obtain

$$H_2\big(z/g(w), w\big) - z g'(w) \{g(w)\}^{-2} H_1\big(z/g(w), w\big) = 0,$$

which is to say

$$\frac{\partial}{\partial w} H\big(z/g(w), w\big) = 0.$$

Thus $H\{z/g(w), w\}$ is independent of w, and so has the form $\Psi(z)$ for some Ψ. In other words,

$$K(z, w) = A(w) H(z, w) = A(w) \Psi\big(z g(w)\big).$$

Theorem 6.5 can be expressed more directly in terms of the $p_n(z)$ as follows.

Theorem 6.9. *If $p_n(z)$ is a polynomial of degree n, the sequence $\{p_n(z)\}$ has a generalized Appell representation with $g_1 = 1$ if and only if there are two sequences of numbers $\{c_n\}$ and $\{b_n\}$ such that*

$$\left. \begin{aligned} z^{n+1} \frac{d}{dz} \{z^{-n} p_n(z)\} + U_0[p_{n-1}(z)] + \\ + U_1[p_{n-2}(z)] + \cdots + U_{n-1}[p_0(z)] = 0, \end{aligned} \right\} \quad (6.10)$$

where U_k is the linear differential operator

$$U_k = c_k + z b_{k+1} \frac{d}{dz}.$$

More explicitly, (6.10) is

$$z^{n+1} [p_n(z) z^{-n}]' = -\sum_{k=0}^{n-1} c_{n-k-1} p_k(z) - z \sum_{k=1}^{n-1} b_{n-k} p_k'(z). \qquad (6.11)$$

Evidently (6.11) can be used to calculate the $p_n(z)$ recursively.

To establish Theorem 6.9 we merely observe that (6.11) is equivalent to (6.7). In fact, if $K(z, w) = \sum p_n(z) w^n$, (6.7) is

$$\sum p_n(z) n w^{n-1} = c(w) \sum p_n(z) w^n + z w^{-1} b(w) \sum p_n'(z) w^n, \qquad (6.12)$$

and (6.11) follows if we express $c(w)$ and $b(w)$ by (6.6) and equate coefficients of w^n. Conversely, if (6.11) holds, (6.12) follows and $K(z, w)$ satisfies (6.7).

Since (6.10) is independent of Ψ, we cannot specialize Theorem 6.9 to get a characterization of Appell polynomials. However, there is a considerable simplification for Brenke polynomials, for which $g(w) = w$ and so $g_0 = 1$, $g_n = 0$ for $n > 1$: a set $\{p_n\}$ has a generating relation

$$A(w)\,\Psi(z\,w) = \sum_{n=0}^{\infty} p_n(z)\,w^n$$

if and only if there are constants c_k such that

$$[z^{-n-1}\,p_{n+1}(z)]' = z^{-n-2}\sum_{k=0}^{n} p_k(z)\,c_{n-k}.$$

In our applications we shall suppose that $A(w)$, $\Psi(t)$, $g(w)$ are all regular at 0.

Chapter II

Representation of entire functions

§ 7. General theory

In this chapter we consider the representation of entire functions by series of generalized Appell polynomials. First we shall see how the class of functions that can be represented, and the number of expansions of a given function, depend on properties of the functions A, Ψ and g. Then we shall study the effect of various specializations and finally we shall illustrate some points of the theory by means of particular sets of polynomials. We suppose throughout the chapter that the function Ψ of (6.1) is a comparison function (§ 2), and hence necessarily entire. We suppose also that $A(w)$ and $g(w)$ are regular at 0. We may then choose a region Ω_w in the w-plane in which $A(w)$ is regular and $g(w)$ is regular and univalent [since we supposed $g'(0) \neq 0$]. If ϱ_0 is the distance from the origin to the nearest point of the boundary of Ω_w, the series (6.1) is convergent for all w in the open disk Δ_w: $|w| < \varrho_0$. Let $\zeta = g(w)$ map Ω_w onto a set Ω_ζ in the ζ-plane, and denote the image of Δ_w by $\Delta_\zeta \subset \Omega_\zeta$. Let the inverse of g be $w = W(\zeta)$, and set $B(\zeta) = A(W(\zeta))$. Then

$$B(\zeta)\,\Psi(z\zeta) = \sum_{n=0}^{\infty} p_n(z)\,\{W(\zeta)\}^n, \qquad (7.1)$$

with the series converging uniformly in compact subsets of Δ_ζ. A set of generalized Appell polynomials is often defined by a generating relation of the form (7.1) instead of (6.1). This is particularly true when the coefficient functionals $\mathcal{L}_n$, rather than the polynomials p_n, are of central interest, as in an interpolation series: cf. § 10 (vi, ix).

Let C be a compact subset of Δ_ζ, and denote by $\Re_\Psi[C]$ the class of entire functions of finite Ψ-type with $D(f) \subset C$ (notation as in § 2). Let Γ be a simple closed contour lying in Δ_ζ and enclosing C but passing through no zero of $B(\zeta)$. Then Γ is an admissible path for (7.1), since

$$\sum_{n=0}^{\infty} p_n(z) \{W(\zeta)\}^n/B(\zeta)$$

converges to $\Psi(z\zeta)$, uniformly on Γ, and we obtain at once:

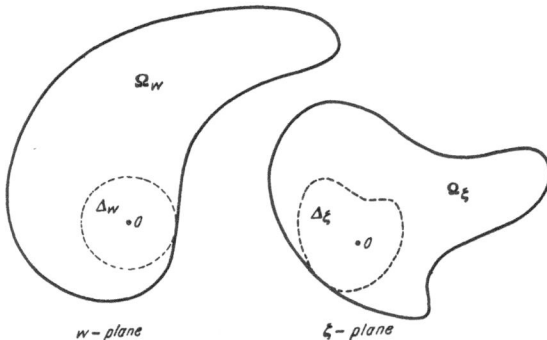

w - plane ζ- plane

Fig. 2. The sets Δ and Ω

Theorem 7.2. *Any f in $\Re_\Psi[C]$ has the convergent expansion*

$$f(z) = \sum_{n=0}^{\infty} \mathscr{L}_n(f)\, p_n(z), \tag{7.3}$$

where

$$\mathscr{L}_n(f) = \frac{1}{2\pi i} \int_\Gamma \frac{\{W(\zeta)\}^n}{B(\zeta)} F(\zeta)\, d\zeta, \tag{7.4}$$

and F is the transform of f defined by means of $\Psi(z\zeta)$ as in (2.4), (2.6).

Corollary. *Any f in $\Re_\Psi[\Delta_\zeta]$ has at least one convergent expansion* $\sum c_n p_n(z)$.

We observe that each choice of an admissible path Γ defines an expansion formula for elements of $\Re_\Psi[C]$. With notation as in § 1, let T_Γ be the transformation, with domain $\Re_\Psi[C]$, sending f into the coefficient sequence $\{\mathscr{L}_n(f)\}$ given by (7.4). If Γ_1 and Γ_2 are two admissible paths which enclose the same zeros of $B(\zeta)$, then $T_{\Gamma_1} = T_{\Gamma_2}$ on $\Re_\Psi[C]$ and Γ_1 and Γ yield the same expansion formula. On the other hand, if Γ_1 encloses a zero $\zeta = \zeta_0$ of $B(\zeta)$ which Γ_2 does not, there are functions f in $\Re_\Psi[C]$ for which $T_{\Gamma_1}(f) \neq T_{\Gamma_2}(f)$. In this case there are at least two different expansion formulas (in the sense of § 1), $f = \sum \mathscr{L}_n(f) p_n(z)$, for functions f in $\Re_\Psi[C]$.

We shall discuss this multiplicity of representations in more detail in § 8. At present, we are content to summarize the observations above as follows.

Theorem 7.5. *At least as many independent expansion formulas apply to the class $\Re_\Psi[C]$ as there are zeros of $B(\zeta)$ in $\Delta_\zeta - C$.*

The expansion formula applying in this case to the *smallest* class of entire functions is that which arises from the choice of Γ as a path enclosing no zeros of $B(\zeta)$ in $\varDelta_\zeta$. As might be expected, the series obtained from this expansion formula is the Whittaker basic series whenever the series for the coefficients (1.4) converge. To verify this, we start from the series

$$\Psi(z\zeta) = \sum_{n=0}^{\infty} p_n(z) \{W(\zeta)\}^n/B(\zeta) = \sum_{n=0}^{\infty} p_n(z) u_n(\zeta),$$

convergent on Γ, where

$$u_n(\zeta) = \sum_{k=n}^{\infty} u_{nk} \zeta^k;$$

the series converges for sufficiently small $|\zeta|$, and is summable on Γ. With $\Psi(t) = \sum_{k=0}^{\infty} \Psi_k t^k$, this leads to

$$\sum_{k=0}^{\infty} \Psi_k z^k \zeta^k = \sum_{n=0}^{\infty} p_n(z) \sum_{k=n}^{\infty} u_{nk} \zeta^k$$

$$= \sum_{k=0}^{\infty} \sum_{n=0}^{k} u_{nk} p_n(z) \zeta^k.$$

Comparing coefficients of ζ^k, we have

$$\Psi_k z^k = \sum_{n=0}^{k} u_{nk} p_n(z).$$

The matrix (π_{kn}) of the Whittaker theory is then given by

$$\pi_{kn} = \begin{cases} u_{nk}/\Psi_k, & k \geq n; \\ 0, & k < n. \end{cases}$$

Let C be a compact subset of $\varDelta_\zeta$ in which $B(\zeta)$ has no zeros and let C_0 be the largest disk, center at 0, contained in C. If $f(z) = \sum_{k=0}^{\infty} f_k z^k$, we have $F(\zeta) = \sum_{k=0}^{\infty} (f_k/\Psi_k) \zeta^{-k-1}$, convergent on Γ if $f \in K[C_0]$ and summable if $f \in K[C]$. Then

$$u_n(\zeta) F(\zeta) = \sum_{k=0}^{\infty} (f_k/\Psi_k) \zeta^{-k-1} \sum_{m=n}^{\infty} u_{nm} \zeta^m,$$

$$\mathscr{L}_n(f) = (2\pi i)^{-1} \int_\Gamma u_n(\zeta) F(\zeta) \, d\zeta = \sum_{k=n}^{\infty} (f_k/\Psi_k) u_{nk}$$

$$= \sum_{k=n}^{\infty} f_k \pi_{kn} = \sum_{k=n}^{\infty} \pi_{kn} f^{(k)}(0)/k!,$$

and this is (1.4) when the last series converges, as it must if $f \in K[C_0]$; when the series diverges it is still summable provided that $f \in K[C]$. We may state the conclusion as a formal theorem.

Theorem 7.6. *With notation as above, if $f \in K_\Psi[C]$ and $B(\zeta)$ has no zeros in the closure of C, then $f(z)$ is represented by its basic series $\sum c_n p_n(z)$ whenever the coefficient formula (1.4) converges; (1.4) is convergent whenever $f \in K_\Psi[C_0]$, and is Mittag-Leffler summable in any case.*

The extension to summable expansions is immediate. We shall give a brief discussion for Mittag-Leffler summability[1]. Let Ω_w^* be the Mittag-Leffler star of Ω_w with respect to the origin, and let Ω_ζ^* be its image under $\zeta = g(w)$. The series (6.1) is Mittag-Leffler summable in

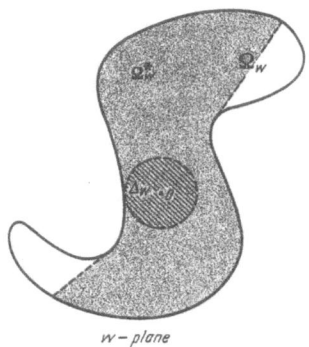 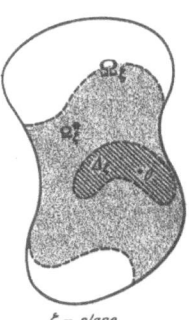

w - plane

ζ - plane

Fig. 3. Star sets

Ω_ζ^*, so that (7.1) is Mittag-Leffler summable in Ω_ζ^*, uniformly in compact subsets. Let C^* be such a compact set and let Γ be a contour lying in Ω_ζ^* and enclosing C^*.

Theorem 7.7. *Any f in $\Re_\psi[C^*]$ has the Mittag-Leffler summable expansion* (7.3) *with coefficients* (7.4), *and any f in $\Re_\psi[\Omega_\zeta^*]$ has at least one Mittag-Leffler summable expansion* $\sum c_n p_n(z)$.

As before, the zeros of $B(\zeta)$ lying in Ω_ζ^* and outside C^* correspond to independent expansion formulas, applying to a given class of functions and valid with Mittag-Leffler summability instead of convergence. We may expect that there will be more of these, since $B(\zeta)$ will often have zeros outside Δ_ζ but in Ω_ζ^*.

§ 8. Multiple expansions

In the preceding section we saw that zeros of $A(w)$ give rise to multiple expansions. Here we shall investigate multiple expansions in more detail; we shall find that they may also arise from non-univalence of $g(w)$, but that if their coefficients grow sufficiently slowly, they can arise only from zeros of $A(w)$.

If a function f has two distinct representations as a series of the form $\sum c_n p_n(z)$, their difference gives rise to a series of the form

$$0 = \sum c_n p_n(z) \tag{8.1}$$

where not all the coefficients are zero. We shall call a series of the

[1] Cf. § 3, p. 12, and BUCK [2], [3].

form (8.1) a *representation of zero*. Conversely, any nontrivial representation of zero allows one to construct an infinity of different expansion formulas. In fact, let

$$f(z) = \sum \mathscr{L}_n(f)\, p_n(z) \tag{8.2}$$

be a given expansion formula, applicable to a class $\mathfrak{R}$ of functions. Let Φ be any linear functional whose domain includes $\mathfrak{R}$, and define a new sequence of functionals by

$$\mathscr{L}_n^*(f) = \mathscr{L}_n(f) + c_n\, \Phi(f).$$

Then, for every $f \in \mathfrak{R}$, we have

$$f(z) = \sum \mathscr{L}_n^*(f)\, p_n(z), \tag{8.3}$$

and if Φ is not identically zero on $\mathfrak{R}$, there is some function in $\mathfrak{R}$ for which (8.2) and (8.3) are different expansions.

We have already observed that some representations of zero arise from zeros of $A(w)$. We can go beyond what follows from Theorem 7.5 by observing in the first place that zeros of $A(w)$ contribute representations of zero according to their multiplicities.

Theorem 8.4. *Suppose that $\{p_n\}$ is a set of generalized Appell polynomials whose generating relation is*

$$A(w)\, \Psi\big(z\, g(w)\big) = \sum p_n(z)\, w^n, \tag{8.5}$$

with $A(w)$ and $g(w)$ regular in a disk Λ with center at 0 that contains a region of univalence Ω_w of $g(w)$. Then each zero of $A(w)$ in Λ gives rise to a different representation of zero, with a k-fold zero contributing k different representations.

For, if α is a zero of $A(w)$ in Λ, then

$$0 = A(\alpha)\, \Psi\big(z\, g(\alpha)\big) = \sum p_n(z)\, \alpha^n. \tag{8.6}$$

If α is at least a double zero, differentiation of (8.5) with respect to w gives

$$0 = A(\alpha)\, z\, g'(\alpha)\, \Psi'\big(z\, g(\alpha)\big) + A'(\alpha)\, \Psi\big(z\, g(\alpha)\big) = \sum p_n(z)\, n\, \alpha^{n-1},$$

which is different from any of the representations (8.6) since $\alpha \neq 0$. If α is a zero of still higher order, repeated differentiation leads to further representations of zero. These representations of zero are convergent; if Λ is not restricted to be a disk, there may also be summable expansions of zero.

However, representations of zero may arise in still another way.

Theorem 8.7. *With notation as in Theorem 8.4, suppose that $g(w)$ is not univalent throughout Λ. Then each pair (α, β) of points of Λ such* *that* $g(\alpha) = g(\beta)$ *but* $A(\alpha) \neq 0$ *and* $A(\beta) \neq 0$ *gives rise to a different representation of zero.*

In fact, we have

$$A(\alpha) \Psi(z g(\alpha)) = \sum \alpha^n p_n(z),$$
$$A(\beta) \Psi(z g(\beta)) = \sum \beta^n p_n(z),$$

and so

$$0 = \sum_{n=0}^{\infty} \left\{ \frac{\alpha^n}{A(\alpha)} - \frac{\beta^n}{A(\beta)} \right\} p_n(z).$$

We can obtain further representations of zero by differentiation if $g'(\alpha) = g'(\beta)$ and $g(\alpha) = g(\beta)$, or if $g'(\alpha) = 0$; still more if there are points α, β, γ of Λ such that $g(\alpha) = g(\beta) = g(\gamma)$; and so on.

It is natural to ask whether there is a characterization of the class of all representations of zero. We shall give a partial answer in the next theorem. If $p_n(z)$ are defined by (8.5), then unless $A(w) \Psi(z g(w))$ is entire we have $\lim \sup |p_n(z)|^{1/n} > 0$, so that $|p_n(z)| > R^{-n}$ for some R and an infinity of n. Consequently it seems not unreasonable to consider the class of representations $0 = \sum c_n p_n(z)$ with $c_n = O(R^n)$ for some R.

Theorem 8.8. *With notation as in Theorem 8.4, let*

$$0 = \sum h_n p_n(z) \tag{8.9}$$

for all z (uniformly in compact sets), where $h_n = O(R^n)$ for some finite R. Let the region Ω_w [in which $g(w)$ is univalent] contain the disk $|w| \leq R$. Then (8.9) is a finite linear combination of the representations which arise as in Theorem 8.4 from the zeros of $A(w)$ in Ω_w.

Theorem 8.8 applies to all Brenke sets [where $g(w) = w$] and in particular to all Appell sets.

To prove Theorem 8.8 let $H(w) = \sum\limits_{n=0}^{\infty} h_n w^{-n-1}$. By assumption, we can choose a contour Γ lying in Ω_w on which both this series and (8.5) converge uniformly. It follows at once that

$$\int_{\Gamma} A(w) \Psi(z g(w)) H(w) \, dw = \sum h_n p_n(z) = 0, \tag{8.10}$$

uniformly on compact sets. If we differentiate the left-hand side of (8.10) repeatedly (with respect to z) and set $z = 0$, we obtain

$$\int_{\Gamma} A(w) H(w) \{g(w)\}^k \, dw = 0, \quad k = 0, 1, 2, \ldots. \tag{8.11}$$

Let $\mathfrak{B}_g$ be the class of functions of the form $f(w) = S\{g(w)\}$, with S analytic on $g(\Omega_w) = \Omega_\zeta$. Then (8.11) implies that

$$\int_{\Gamma} A(w) H(w) f(w) \, dw = 0 \tag{8.12}$$

for all $f \in \mathfrak{B}_g$. Since g was assumed univalent on Ω_w, we can take S as, in particular, g^{-1}, and so $\mathfrak{B}_g$ contains all the functions w^k, $k = 0, 1, 2, \ldots$. In particular, then,

$$\int_\Gamma A(w) H(w) w^k dw = 0, \quad k = 0, 1, 2, \ldots,$$

and this implies that[1] $A(w) H(w) = a(w)$, say, is analytic inside Γ. Then

$$H(w) = \sum h_n w^{-n-1} = a(w)/A(w)$$

inside Γ. If $A(w)$ has no zeros inside Γ, it follows that $H(w)$ is analytic for all w, and hence is constant; since $H(\infty) = 0$, $H(w) \equiv 0$ and all h_n are zero. In this case, therefore, 0 has only the trivial representation [at least among representations (8.9) with $h_n = O(R^n)$].

If $A(w)$ has zeros α_k (simple or multiple) inside Γ (necessarily finite in number) $H(w)$ is rational with poles α_k. Writing the partial fraction decomposition of $H(w)$, we find that the h_n are linear combinations of the coefficients of $(w - \alpha_k)^{-r}$ $(r = 1, 2, \ldots, r_k)$ in its Laurent series.

To obtain the additional representations of zero described in Theorem 8.7, we must allow the h_n to increase more rapidly. Suppose that $h_n = O(R^n)$, where the disk $|w| \leq R$ is now allowed to extend beyond Ω_w, but must still lie in Λ. Let Γ'' be a contour in Λ but outside this disk, and repeat the argument. We find again that

$$\int_{\Gamma''} A(w) H(w) f(w) dw = 0 \qquad (8.13)$$

for all $f \in \mathfrak{B}_g$. Now, however, $\mathfrak{B}_g$ is a much restricted subspace of the class of functions analytic in Λ, and it does *not* follow that $A(w) H(w)$ is analytic in Λ. In particular, since g is not univalent in Λ we can choose two points α and β such that $g(\alpha) = g(\beta)$. Then we must have $f(\alpha) = f(\beta)$ for all $f \in \mathfrak{B}_g$, and hence

$$\int_{\Gamma''} \left(\frac{1}{w - \alpha} - \frac{1}{w - \beta} \right) f(w) dw = 0$$

for all $f \in \mathfrak{B}_g$. This shows that $H(w)$ may have poles at points in Λ which are not zeros of $A(w)$, and these give rise to representations of zero which are different from those described in Theorem 8.8—in fact, the representations obtained in Theorem 8.7.

It would be interesting to have a simple characterization of the functions $F(w)$ $[= A(w) H(w)]$ that are analytic on Γ'' and obey

$$\int_{\Gamma''} F(w) f(w) dw = 0$$

for all $f \in \mathfrak{B}_g$.

---

[1] That (8.10) implies this conclusion is a generalization of a lemma of Pólya's (see Boas [3], p. 110) for $\Psi(t) = e^t$. Our proof is different.

§ 9. Appell polynomials

We shall illustrate our general theory first by applying it to some more or less specialized classes of polynomial sets and then, by specializing still further, to specific polynomial sets. We have tried to select our illustrative sets from among those already discussed in the literature, rather than relying on examples constructed ad hoc. Our collection of examples is far from exhaustive: we have included some sets to illustrate particular points in the general theory, and others to test the power of our methods in dealing with a well-known set. Our first specialization is to the class of Appell polynomials (APPELL [1], BOURBAKI [1]). These are obtained by taking $g(w) = w$ and $\Psi(t) = e^t$ in (6.1), so that they are defined by

$$A(w)\, e^{zw} = \sum_{n=0}^{\infty} p_n(z)\, w^n, \qquad A(0) \neq 0. \tag{9.1}$$

In this case the result described in § 7 can be summarized as follows (MARTIN [1]).

Theorem 9.2. *If $p_n(z)$ are the Appell polynomials defined by (9.1), if $A(w)$ is regular in the region Ω, and if Δ is the largest circular disk, with center at 0, in Ω, and Δ has radius ϱ_0, then every entire function of exponential type $\tau < \varrho_0$ has the representation*

$$f(z) = \sum_{n=0}^{\infty} \mathscr{L}_n(f)\, p_n(z),$$

where

$$\mathscr{L}_n(f) = \frac{1}{2\pi i} \int_{\Gamma} \frac{w^n}{A(w)}\, F(w)\, dw, \tag{9.3}$$

$F(w)$ is the Borel transform of $f(z)$, and Γ is a circumference $|w| = \varrho$ with $\tau < \varrho < \varrho_0$ on which $A(w) \neq 0$.

The zeros of $A(w)$ in Δ give rise to different expansion formulas. The expansion reduces to the basic series when $A(w)$ has no zeros inside Γ. Hence, the critical type for the basic series expansion is the modulus of the zero of $A(w)$ that is closest to the origin.

With any set of polynomials we can associate another which we call the reversed set, for lack of a generally accepted term.

Definition. *If $P(z) = a_0 + a_1 z + \cdots + a_n z^n$, $a_n \neq 0$, the reverse of P is the polynomial $P^*(z) = z^n P(1/z) = a_0 z^n + a_1 z^{n-1} + \cdots + a_{n-1} z + a_n$.*

We note here a characterization of Appell polynomials that is equivalent to the explicit formula quoted in § 6.

Theorem 9.4. *The Appell polynomials defined by (9.1) are the inverse Laplace transforms of the functions $t^{-1} A_n^*(t^{-1})$, where $A_n(w)$ are the partial sums of the power series of $A(w)$, and A_n^* is the reverse of A_n.*

For, the Laplace transform of $A(w) e^{zw}$, as a function of z, is $A(w)(t-w)^{-1} = t^{-1} A(w) \sum\limits_{n=0}^{\infty} t^{-n} w^n$. If $A(w) = \sum\limits_{n=0}^{\infty} a_n w^n$, this is

$$t^{-1} \sum_{n=0}^{\infty} w^n \sum_{k=0}^{n} a_k t^{k-n} = \sum_{n=0}^{\infty} w^n t^{-1} A_n^*(t^{-1}).$$

We now consider some particular sets of Appell polynomials[1].

(i) **Bernoulli polynomials and generalizations.** (Effect of singularities of A.) Bernoulli polynomials furnish one of the best known examples of an Appell set. They arise from the choice $A(w) = w/(e^w - 1)$:

$$\frac{w}{e^w - 1} e^{zw} = \sum_{n=0}^{\infty} B_n(z) w^n/n! = \sum_{n=0}^{\infty} p_n(z) w^n.$$

Here $A(w)$ has singular points at $w = \pm 2k\pi i$, and $A(w)$ has no zeros. There is, then, only one expansion formula whose coefficients are $O(R^n)$ for some $R < 2\pi$; these coefficients are

$$\mathcal{L}_n(f) = (2\pi i)^{-1} \int_\Gamma w^{n-1}(e^w - 1) F(w)\, dw = \begin{cases} \int_0^1 f(x)\, dx, & n = 0; \\ f^{(n-1)}(1) - f^{(n-1)}(0), & n > 0. \end{cases}$$

Since[2]

$$p_n(z) = -\frac{e^{2\pi i z} + (-1)^n e^{-2\pi i z}}{(2\pi i)^n} + O\left(\frac{e^{3\pi |z|}}{(3\pi)^{n-1}}\right),$$

we have $\liminf |p_n(z)|^{1/n} = 1/(2\pi)$, and there cannot therefore be any convergent expansions $0 = \sum c_n p_n(z)$ with $\limsup |c_n|^{1/n} > 2\pi$. The possibility of a representation of zero with $c_n = O((2\pi)^n)$ is not excluded by this discussion.

Any entire function of exponential type less than 2π has the convergent expansion

$$f(z) = \int_0^1 f(x)\, dx + \sum_{n=1}^{\infty} \{f^{(n-1)}(1) - f^{(n-1)}(0)\} B_n(z)/n!. \qquad (9.5)$$

If the conjugate indicator diagram of $f(z)$ contains no point iy with $|y| \geq 2\pi$, the series (9.5) is Mittag-Leffler summable to $f(z)$. The example $f(z) = \sin 2\pi z$ shows that the number 2π cannot be replaced by anything larger. The series in (9.5) is equivalent to the familiar Euler-Maclaurin summation formula, which we have now shown to converge for all functions of exponential type less than 2π, and to be Borel (or Mittag-Leffler) summable if f is of finite exponential type

---

[1] Most of the special sets of polynomials that we consider appear in ERDÉLYI [1], where references to the original sources are given. We usually refer only to this source.

[2] See, e.g., WHITTAKER [1], p. 21.

and obeys $h\left(\pm \frac{\pi}{2}; f\right) < 2\pi$. This type restriction can be overcome in much the same way that we used in § 4 for the Lidstone series. Computing the residue of $w\,e^{zw}/(e^w - 1)$ at the poles $\pm 2\pi i$, we see that

$$\frac{w\,e^{zw}}{e^w - 1} + \frac{4\pi}{w^2 + 4\pi^2}\{w\sin(2\pi z) + 2\pi\cos(2\pi z)\}$$

is regular for all w with $|w| < 4\pi$. Repeating this for the additional poles $\pm 2k\pi i$, we find that

$$\frac{w\,e^{zw}}{e^w - 1} = \sum_1^N \frac{-4\pi k w}{w^2 + 4\pi^2 k^2}\sin(2\pi k z) + \sum_1^N \frac{-8\pi^2 k^2}{w^2 + 4\pi^2 k^2}\cos(2\pi k z) + G(z, w),$$

where $G(z, w) = \sum q_n(z)\,w^n$ converges for $|w| < 2(N+1)\pi$. The coefficient functions $q_n(z)$ are formed by adding certain trigonometric polynomials to the Bernoulli polynomials. If we multiply both sides of this expansion by $(e^w - 1)/w$, we obtain an expansion of the Pólya kernel e^{zw}. The usual routine therefore leads to the following result: *any entire function f of finite exponential type τ has a convergent expansion in the form*

$$f(z) = \int_0^1 f(t)\,dt + \sum_1^\infty q_n(z)\,[f^{(n-1)}(1) - f^{(n-1)}(0)] +$$

$$+ \sum_1^N \{A_k \sin(2\pi k z) + C_k \cos(2\pi k z)\}$$

where $N = [\tau/(2\pi)]$. The coefficients A_k and C_k can be computed from the Borel transform $F(w)$ of f.

The *generalized Eulerian polynomials*[1] have $A(w) = (1 - \alpha)^\lambda/(1 - \alpha e^w)^\lambda$, $\alpha \neq 0, 1$ (these are the usual Euler polynomials when $\alpha = -1$ and $\lambda = 1$). Here $A(w)$ has singular points at all determinations of $-\log \alpha$. The results are similar to those for the Bernoulli polynomials, except that the critical type is now the smallest value of $\{(\log|\alpha|)^2 + (\arg \alpha)^2\}^{\frac{1}{2}}$ instead of 2π (and hence π when $\alpha = -1$).

The *generalized Bernoulli polynomials*[2] have $A(w) = w^l/(e^w - 1)^l$ with positive integral l; their expansion properties are the same as those of the Bernoulli polynomials. There are also *Bernoulli polynomials of higher order*[2], with

$$A(w) = \frac{(e^{\omega_1 w} - 1)\dots(e^{\omega_l w} - 1)}{\omega_1\dots\omega_l\,w^l},$$

or the reciprocal of this. In the second case (positive order) the expansion properties are the same as those of the Bernoulli polynomials except that the critical type for convergence is now given by the modulus of the w closest to the origin for which $\omega_j w = 2\pi i$; and the functions with

---

[1] For various particular cases see ERDÉLYI [1], vol. 3, pp. 252—254; CARLITZ [1]; SUMNER [1], p. 441.

[2] ERDÉLYI [1], vol. 3, pp. 253—254.

Mittag-Leffler summable expansions are those whose conjugate indicator diagrams avoid the boundary of the star domain obtained by deleting from the w-plane the rays containing the solutions of $\omega_j w = 2\pi i$ $(j = 1, 2, \ldots, l)$. In the case displayed above (negative order), the situation is entirely different, since $A(w)$ has an infinity of zeros but no singular points. Thus any entire function f of exponential type has an infinity of distinct convergent expansions in series of Bernoulli polynomials of negative order, and there are infinitely many distinct representations of zero.

(ii) **A set of Laguerre polynomials.** These are the polynomials considered in § 5, an Appell set with $A(w) = (1 - w)^\lambda$; it turns out that $p_n(z) = (-1)^n L_n^{(\lambda-n)}(z)$. The "Boole polynomials of the second kind" considered by PETERS [2] have $A(w) = (1 + \frac{1}{2}w)^\lambda$ and are expressible in terms of the present set by a change of variable.

When λ is a positive integer the zero of $A(w)$ at $w = 1$ gives rise to a non-trivial representation of zero.

When λ is not a nonnegative integer, Ω can be taken to be the plane with a cut from 1 to ∞; Δ is the disk $|w| < 1$; and $\varrho_0 = 1$. Since $A(w)$ has no zeros in Ω, the basic series represents all entire functions of exponential type less than 1. There is no nontrivial representation of zero whose coefficients are $O(R^n)$, $R < 1$. SHEFFER [1] has used the case $\lambda = -1$. Here the $p_n(z)$ are the *partial sums of the exponential series* and $\mathscr{L}_n(f) = f^{(n)}(0) - f^{(n+1)}(0)$. The expansion of any entire function of exponential type is Mittag-Leffler summable if the conjugate indicator diagram of the function avoids the point -1; for example, any trigonometric polynomial has a Mittag-Leffler summable expansion.

(iii) **Hermite polynomials.** (Entire A with no zeros.) These familiar polynomials can be defined by

$$\exp(2zt - t^2) = \sum_{n=0}^\infty H_n(z)\, t^n/n!.$$

If we put $t = w/2$ this becomes

$$e^{-w^2/4}\, e^{zw} = \sum_{n=0}^\infty \frac{H_n(z)}{2^n n!}\, w^n,$$

so that $H_n(z)$ are multiples of the Appell polynomials corresponding to $A(w) = e^{-w^2/4}$. Since this is entire and has no zeros, every entire function f of exponential type can be represented by a convergent series of Hermite polynomials, in fact by its basic series. In comparing this result with the literature, we must observe that what are usually considered are expansions in terms of the orthogonal functions $e^{-z^2/2} H_n(z)$. Thus we have shown that $f(z)$ can be represented as a series of these Hermite functions if $f(z) e^{z^2/2}$ is an entire function of exponential type. Expansion theorems for more general classes of functions are known

(HILLE [1]); for an account of them see § 18. There are no non-trivial representations of zero with coefficients $O(R^n)$ for some finite R.

We defined the reversed set of a set of polynomials earlier in this section. The reversed set of a set of Brenke polynomials (and in particular of a set of Appell polynomials) is a set of the same kind provided that $A(w)$ has no zero coefficients. Since we shall use this fact several times, we give it a formal statement.

Lemma 9.6. *Let* $\{p_n(z)\}$ *be defined by*

$$A(w)\,\Psi(z\,w) = \sum_{n=0}^{\infty} p_n(z)\,w^n, \qquad (9.7)$$

and let $p_n^*(z) = z^n p_n(1/z)$. *Then*

$$\Psi(w)\,A(z\,w) = \sum_{n=0}^{\infty} p_n^*(z)\,w^n.$$

We have only to replace z by $1/z$, and then w by zw, in (9.7).

We give two illustrations of the use of this principle.

(iv) Reversed Laguerre polynomials. (Entire A with an infinity of zeros.) A generating relation for the generalized Laguerre polynomials is[1]

$$e^w (w\,z)^{-\lambda/2} J_\lambda\left(2(w\,z)^{\frac12}\right) = \sum_{n=0}^{\infty} L_n^{(\lambda)}(z)\,w^n/(n+\lambda)!. \qquad (9.8)$$

Applying Lemma 9.6, we see that the reverses of $L_n^{(\lambda)}(z)$ are generated by

$$e^{z\,w}\,w^{-\lambda/2} J_\lambda(2\,w^{\frac12}) = \sum_{n=0}^{\infty} z^n\,L_n^{(\lambda)}(1/z)\,\frac{w^n}{(n+\lambda)!}\,.$$

Thus the reversed Laguerre polynomials are Appell polynomials corresponding to $A(w) = w^{-\lambda/2} J_\lambda(2\,w^{\frac12})$. This is entire and has infinitely many zeros, so that every entire function f of exponential type can be expanded in infinitely many different series of reversed Laguerre polynomials; there are an infinity of distinct representations of zero corresponding to the zeros of $J_\lambda(2\,w^{\frac12})$.

The expansion properties of the reversed set tell us nothing about the expansion properties of the set itself. For expansion properties of the Laguerre polynomials see below, example (x).

(v) Reversed Rainville polynomials. The set

$$R_n(z) = (1-z^2)^{n/2}\,P_n\{(1-z^2)^{-\frac12}\} = \frac{1}{\pi}\int_0^\pi (1+z\cos\theta)^n\,d\theta,$$

where P_n denotes a Legendre polynomial, has the generating function[2]

$$e^w I_0(z\,w) = \sum_{n=0}^{\infty} R_n(z)\,w^n/n!.$$

---

[1] ERDÉLYI [1], vol. 3, p. 262 (9).
[2] ERDÉLYI [1], vol. 3, p. 262 (5); RAINVILLE [2].

The $R_n(z)$ are not generalized Appell polynomials since $I_0(w)$ is even. Indeed, the definition of $R_n(z)$ shows that they are all even functions. However, if we apply the reasoning of Lemma 9.6 we get

$$I_0(w)\, e^{zw} = \sum_{n=0}^{\infty} R_n^*(z)\, w^n/n!\,.$$

Thus the $R_n^*(z)$ are Appell polynomials. Since $I_0(w)$ is entire and has infinitely many zeros, we have the same conclusions as in (iv).

The Rainville polynomials themselves are considered in example (xx), § 11.

§ 10. Sheffer polynomials

A class more general than the class of Appell polynomials is obtained by taking $\Psi(t) = e^t$ in (6.1). The generating relation is now

$$K(z,w) \equiv A(w)\, e^{z\, g(w)} = \sum_{n=0}^{\infty} p_n(z)\, w^n. \tag{10.1}$$

These polynomials are called sets of zero type by SHEFFER [2], [3], poweroids by STEFFENSEN [1], [2], and sets of generalized Appell type by the Bateman Manuscript Project (ERDÉLYI [1]).

Since we require that $A(0) \neq 0$, (10.1) can also be written with kernel $K(z; w) = \exp\{a(w) + z\, g(w)\}$, where $a(w)$ and $g(w)$ are both regular at 0, $g(0) = 0$, $g'(0) \neq 0$. In this connection, the following result is of interest (BOAS and BUCK [1]).

Theorem 10.2. *The only kernels of the form $K(z, w) = H\{a(w) + z\, g(w)\}$ with $H(0) = 1$ that generate generalized Appell polynomials are those with $H(t) = e^t$ or $H(t) = (1 - t)^\lambda$.*

To prove this, differentiate $K(z, w) = A(w)\, \Psi(z\, g(w))$ with respect to z to obtain

$$K(0, w) = A(w)\, \Psi(0)\,,$$
$$K_z(0, w) = A(w)\, \Psi'(0)\, g(w)\,,$$
$$K_{zz}(0, w) = A(w)\, \Psi''(0)\, \{g(w)\}^2\,.$$

Thus a necessary condition for $K(z, w)$ to generate generalized Appell polynomials is that, for $z = 0$ and w in a neighborhood of 0,

$$K K_{zz} = k (K_z)^2, \tag{10.3}$$

where k is a constant. If we apply this necessary condition to $K(z, w) = H\{a(w) + z\, g(w)\}$ we find

$$K(0, w) = H\big(a(w)\big)\,,$$
$$K_z(0, w) = H'\big(a(w)\big)\, g(w)\,,$$
$$K_{zz}(0, w) = H''\big(a(w)\big)\, \{g(w)\}^2\,,$$
$$H\big(a(w)\big)\, H''\big(a(w)\big)\, \{g(w)\}^2 = k\, [H'\big(a(w)\big)]^2 \{g(w)\}^2. \tag{10.4}$$

Solving (10.4) with $H(0)=1$ we obtain $H(t)=e^{\gamma t}$ or $H(t)=(1+\gamma t)^{\lambda}$. Since the constant γ can be absorbed into the functions $a(w)$ and $g(w)$, we may take $\gamma=1$ or $\gamma=-1$, respectively.

In this section we study generalized Appell polynomials generated by kernels of the first form in the conclusion of Theorem 10.2. Those generated by kernels of the second form will be discussed in §§ 14, 15.

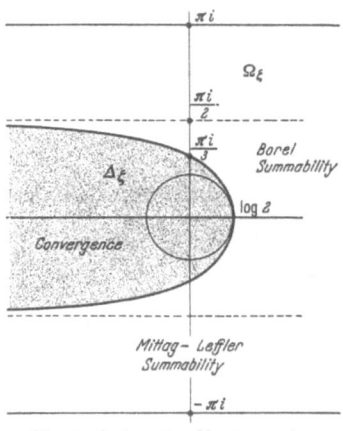

We continue with some specific examples.

(vi) **General difference polynomials.** These are obtained by taking $A(w)=1$, $g(w)=\zeta$, where $w=(e^{\zeta}-1)e^{\beta\zeta}$; that is, they are defined by the generating relation

$$e^{z\zeta}=\sum_{n=0}^{\infty}p_n(z)\{(e^{\zeta}-1)e^{\beta\zeta}\}^n.$$

For $\beta=1$ they are *Selberg's polynomials* (SELBERG [1]); for $\beta=0$, the *Newton* (or *binomial*) *polynomials*; for $\beta=-\frac{1}{2}$, the *Stirling interpolation polynomials*. They have all been studied in detail elsewhere[1]. We quote the principal results. The Newton polynomials have $g(w)=\log(1+w)$, and may be written explicitly in the form

$$p_n(z)=\binom{z}{n}=z(z-1)\ldots(z-n+1)/n!.$$

The principal value of the logarithm is to be understood in the definition of $g(w)$, so that $g(0)=0$. We may take Ω_w to be the w-plane with the ray $-\infty<u\leq-1$ deleted; then Ω_{ζ} is the strip $|\Im(\zeta)|<\pi$. The disk Δ_w is $|w|<1$, and Δ_{ζ} is the set lying inside Ω_{ζ} on which $|e^{\zeta}-1|<1$. It follows that every entire function of exponential type whose conjugate indicator diagram is inside Δ_{ζ} can be expanded in a convergent series of Newton polynomials with coefficients $\Delta^n f(0)$; in terms of the indicator $h(\theta)$, this condition is[2]

$$h(\theta)<\cos\theta\log(2\cos\theta)+\theta\sin\theta, \qquad |\theta|<\pi/2.$$

In particular, functions of exponential type less than log 2 can be expanded. There is no nontrivial representation of zero with coefficients $O(R^n)$, $R<1$.

The Borel star (polygon) of Ω_w is the half plane $\Re(w)>-1$ whose image in Ω_{ζ} is the strip $|\Im(\zeta)|<\pi/2$. Hence the Newton interpolation series for f is Borel summable for all entire functions of finite exponential

[1] See, in particular, BUCK [3], where references to earlier work will be found.
[2] PÓLYA and SZEGÖ [1], vol. 1, p. 106, problem 114.

type which obey $h(\pm\pi/2)<\pi/2$. The Mittag-Leffler star of Ω_w is Ω_w itself, so that $\Omega_\zeta^*=\Omega_\zeta$. From Theorem 7.7, we conclude that the Newton series is Mittag-Leffler summable for those f which obey the weaker condition $h(\pm\pi/2)<\pi$. (See Fig. 4.)

Since the coefficient functionals associated with the Newton series are the differences

$$\Delta^n f(0) = (-1)^n \sum_{k=0}^{n}\binom{n}{k}(-1)^k f(k)$$

these results yield a method for computing the values of f in terms of the values $\{f(k)\}$, $k=0, 1, 2, \ldots$. When f is of type less than $\log 2 = 0.693\ldots$, this is effected by a convergent series; but when the type is larger, in general a summability method must be used. Since this is often awkward, the following simple modification has some interest; it is similar in nature to one that was employed in discussing Lidstone series in § 4. Let f be any entire function of exponential type less than $\pi/2=1.57\ldots$. Referring to Fig. 4, we observe that for sufficiently large C, the function $g(z)=e^{-Cz}f(z)$ will have its conjugate indicator diagram $D(g)$ entirely within the convergence region Δ_ζ. In particular, $g(z)$ is the sum of the convergent Newton series for g. We have thus obtained a representation

$$f(z) = e^{Cz}\sum_{n=0}^{\infty}\binom{z}{n}\mathscr{L}_n(f),$$

where

$$\mathscr{L}_n(f) = \Delta^n g(0) = (-1)^n\sum_{k=0}^{n}\binom{n}{k}(-e^{-C})^k f(k).$$

This therefore yields a simple convergent series representation for $f(z)$ in terms of the values $\{f(k)\}$, $k=0, 1, \ldots$, whenever f has type less than $\pi/2$.

We turn now to the difference polynomials with $\beta>0$. The case $\beta=1$ is typical. As an interpolation problem, we seek to express $f(z)$ in terms of the "moving differences" $\mathscr{L}_n(f)=\Delta^n f(n)$. [For general β, these become $\Delta^n f(\beta n)$.] The corresponding polynomial set is

$$p_n(z) = \frac{z}{n}\binom{z-\beta n-1}{n-1}.$$

As shown in Figs. 5, 6, the set Ω_w can be taken as the w-plane, cut along the negative real axis from $-\infty$ to the point $w_0=-\beta^\beta/(1+\beta)^{1+\beta}$. The set Ω_ζ is the set of $\zeta=s+it$ lying in the strip $|t|<\pi/(1+\beta)$, and bounded by the curve whose equation is $s=\log\sin\beta t-\log\sin(1+\beta)t$. The disk Δ_w is {all w with $|w|<|w_0|$}, and Δ_ζ is the subset of Ω_ζ described by $|e^{\beta\zeta}(e^\zeta-1)|<|w_0|$. Specializing to the Selberg case, $\beta=1$, $w_0=-\frac{1}{4}$, and Δ_ζ is a small convex tear-drop shaped set containing the disk $|\zeta|<\log(\frac{1}{2}+\sqrt{\frac{1}{2}})=0.188\ldots$.

3*

Accordingly, any entire function of exponential type $\tau < 0.188$ can be represented by the convergent series $\sum p_n(z)\,\Delta^n f(n)$. For this series to be Mittag-Leffler summable, it is sufficient that the indicator set $D(f)$ lie in Ω_ζ. This holds, for example, if f is of type less than $\log 2$, or more generally, if $h(\theta;f) < (\pi - |\theta|)\,|\sin\theta| - \cos\theta\,\log(-2\cos\theta)$ for $\pi/2 \leq |\theta| \leq \pi$. The Borel star of Ω_w is the half-plane $\Re(w) > w_0$, whose image is an unbounded subset of Ω_ζ having a sharp corner at $\zeta_0 = -\log 2$, and containing the disk $|\zeta| < 0.35$; the Selberg series is therefore Borel summable for functions of type

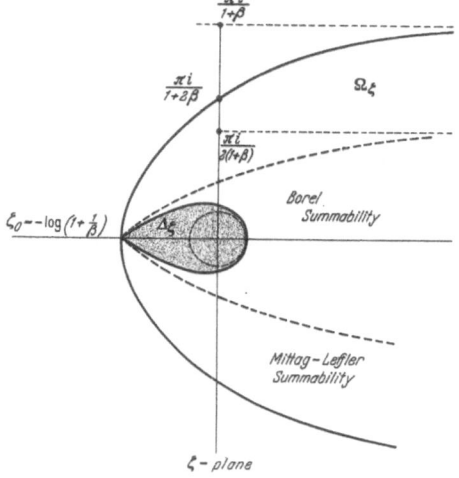

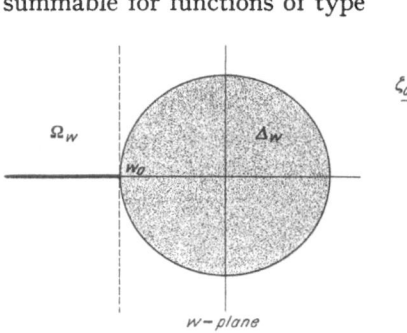

Fig. 5. w-plane for the general difference series with $\beta > 0$　　　Fig. 6. ζ-plane for the general difference series with $\beta > 0$

less than 0.35. For values of β which are positive, but quite small, Fig. 6 is no longer a reliable guide; when $\beta < \tfrac{1}{12}$, the corresponding diagram will give convergence and summability constants which are closer to those for the Newton series, $\beta = 0$.

When β is negative, and lies between 0 and $-\tfrac{1}{2}$, the situation is radically different. The case $\beta = -\tfrac{1}{2}$ yields the Stirling interpolation series. Here the coefficient functionals are the "central differences" $\Delta^n f(-n/2)$. The function $g(w)$ is $2\sinh^{-1}(w/2)$, and Ω_w may be taken as the w-plane with cuts running from $2i$ and $-2i$ to infinity, as shown in Fig. 7. The set Ω_ζ becomes the strip $|\Im(\zeta)| < \pi$. In Ω_w, Δ_w is the disk $|w| < 2$, and its image Δ_ζ is the set of points ζ in Ω_ζ for which $|\sinh(\zeta/2)| < 1$. This is a convex lens-shaped set whose support function is[1]

$$
k(\theta) = \begin{cases}
2\cos\theta\,\log\{\sqrt{2}\cos\theta + \sqrt{\cos 2\theta}\,\} + \\
\qquad + 2\sin\theta\,\sin^{-1}(\sqrt{2}\sin\theta), & |\theta| \leq \pi/4; \\[4pt]
\pi\,|\sin\theta|, & \pi/4 < |\theta| \leq \pi/2; \\[4pt]
k(\theta \pm \pi), & \pi/2 < |\theta| \leq \pi.
\end{cases}
$$

---

[1] See Footnote 2, p. 34.

Accordingly, any entire function of exponential type such that $h(\theta;f)<k(\theta)$ for all θ has a convergent Stirling interpolation series. In particular, since the disk $|\zeta|<\log(3+\sqrt{8})=1.76\ldots$ lies in Δ_ζ, this holds for any function of exponential type less than 1.76. Since the Mittag-Leffler star of Ω_w is Ω_w itself, $\Omega_\zeta^*=\Omega_\zeta$. Accordingly, the Stirling series is ML-summable for any function f which obeys $h(\pm\pi/2;f)<\pi$. The situation with respect to Borel summability is complicated by the fact that the image in Ω_ζ of the strip $|v|<2$, which is the Borel polygon of Ω_w, is not a convex set. It is readily seen that it contains the disk

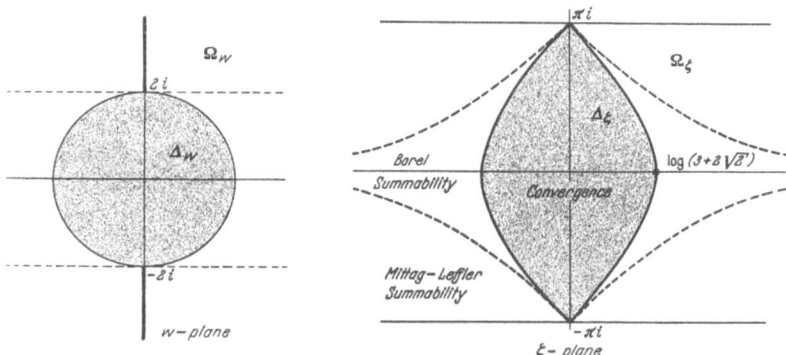

Fig. 7. w-plane for Stirling series   Fig. 8. ζ-plane for Stirling series

$|\zeta|\leqq\pi/\sqrt{2}$, so that the Stirling series is Borel summable for any function of type at most $\pi/\sqrt{2}$. The case when β is a complex number with real part between 0 and $-\frac{1}{2}$ has been studied by R. DeMar[1]. There is no need to study separately the series that result from the choice of $\beta<-\frac{1}{2}$, since these are equivalent to the general difference interpolation series for $\beta>-\frac{1}{2}$, applied to the function $g(z)=f(-z)$.

(vii) Poisson-Charlier, Narumi and Boole polynomials. Consider the generalized Appell polynomials with $g(w)=\log(1+w)$ and, respectively,

$$A(w)=e^w, \tag{10.5}$$

$$A(w)=w^{-k}\{\log(1+w)\}^k, \tag{10.6}$$

$$A(w)=\{1+(1+w)^h\}^\lambda. \tag{10.7}$$

We take the principal values of the logarithms and the powers, and suppose $h>0$. From (10.5) we have the Poisson-Charlier polynomials[2]; from (10.6), except for a shift in index, a set considered by Narumi[2], and from (10.7) the Boole polynomials[3] for $\lambda=1$ and the generalized Boole polynomials of Peters [1] for general λ.

[1] In a Ph. D. thesis (1960), University of Wisconsin; see also DeMar [1], [2], [3].

[2] Erdélyi [1], vol. 3, p. 255 (49); p. 258 (3). Cf. the middle of p. 18.

[3] Jordan [1], pp. 317ff., gives the formal expansion of a function into a series of Boole polynomials.

For (10.5) and (10.6) the expansion properties are exactly the same as for Newton polynomials, since in the first case $A(w)$ is entire and in the second case the only singularity of $A(w)$ is on the boundary of the region Ω_w used for Newton polynomials. For (10.7), $(1+w)^h$ has a singularity on the boundary of Ω_w (or no singular points), and so when λ is a positive integer the expansion properties of the Appell polynomials corresponding to (10.7) are again the same as for Newton polynomials, except that $1+(1+w)^h$ may have zeros inside Ω_w and hence there will be nontrivial representations of zero. This is in particular true for the ordinary Boole polynomials. When λ is not a positive integer, the class of functions that can be represented is generally smaller because of the singular points of A introduced by the zeros of $1+(1+w)^h$. For $0<h\leq 3$ this function has no zeros in Δ_w; for $0<h\leq 1$, it has no zeros in Ω_w. Hence, in particular, the expansion of an entire function in a series of Boole polynomials has the same convergence properties as its Newton series if $0<h\leq 3$, and the same Mittag-Leffler summability properties if $0<h\leq 1$. For $h>3$ we can say that functions of exponential type less than $\log\left(1+2\sin\left(\frac{1}{2}\pi/h\right)\right)$ have convergent expansions, and more precise statements could be made in terms of $h(\theta)$.

(viii) **Mittag-Leffler polynomials**[1]. These are defined by $A(w)=1$, $g(w)=\log\dfrac{1+w}{1-w}$ (with the principal value of the logarithm). We may take Ω_w to be the w-plane, cut from 1 to ∞ and from -1 to $-\infty$ along the real axis. The disk Δ_w is $|w|<1$. The set Ω_ζ is the strip $|\Im(\zeta)|<\pi$, and Δ_ζ is $|\Im(\zeta)|<\pi/2$; Ω_w^* and Ω_ζ^* coincide with Ω_w and Ω_ζ, respectively. The conditions for convergence and Mittag-Leffler summability of the series of Mittag-Leffler polynomials for an entire function of exponential type whose indicator is $h(\theta)$ are then that $h(\pm\pi/2)<\pi/2$ and $h(\pm\pi/2)<\pi$, respectively. *Pidduck's polynomials*[2], which are defined by taking the same $g(w)$, but $A(w)=(1-w)^{-1}$, have the same expansion properties[3].

(ix) **Abel interpolation series.** Here

$$A(w)=1, \qquad g(w)=\zeta, \qquad w=\zeta e^\zeta,$$
$$p_n(z)=z(z-n)^{n-1}/n!, \qquad \mathcal{L}_n(f)=f^{(n)}(n).$$

The expansion theory has been discussed in detail elsewhere[4]; the main results are as follows. We may take Ω_w to be the w-plane with the interval $-\infty<u\leq -1/e$ deleted; then Ω_ζ is bounded by the curve whose polar equation is $\varrho=(\pi-|\varphi|)\csc|\varphi|$. The disk Δ_w is $|w|<1/e$,

[1] ERDÉLYI [1], vol. 3, p. 248 (22).

[2] ERDÉLYI [1], vol. 3, p. 248 (25).

[3] WEISNER [Amer. Math. Monthly **64**, 747, problem 4766 (1957)] considers the closely related case $g(w)=\tan^{-1}w$, $A(w)=(1+w^2)^{-\frac{1}{2}}$.

[4] See, in particular, BUCK [3], where references to earlier work will be found.

and $\varDelta_\zeta$ is defined by $|\zeta\, e^{1+\zeta}| < 1$. Every entire function of exponential type whose conjugate indicator diagram lies in $\varDelta_\zeta$ can be expanded in a convergent Abel series; in particular, this is the case when the function is of type less than the positive root of $xe^{1+x} = 1$ (approximately 0.278). The regions $\varOmega_w^*$ and $\varOmega_\zeta^*$ coincide with $\varOmega_w$ and $\varOmega_\zeta$, respectively. Hence functions of exponential type have Mittag-Leffler summable Abel expansions when their conjugate indicator diagrams lie in $\varOmega_\zeta$, and in particular if they are of type less than 1.

This type limitation is best possible, as is shown by the function $f(z) = z e^{-z}$ for which we have

$$\mathscr{L}_n(f) = f^{(n)}(n) = 0,$$

$$n = 0, 1, 2, \ldots.$$

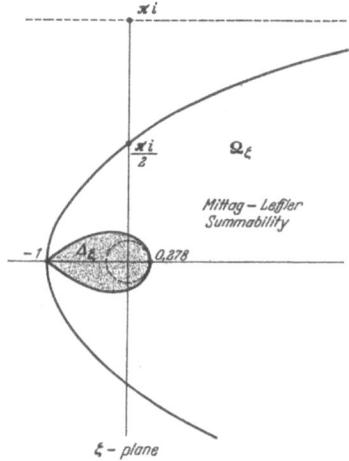

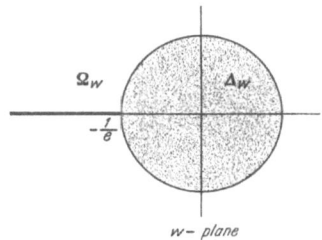

Fig. 9. w-plane for the Abel series   Fig. 10. ζ-plane for the Abel series

Because of the analogy between differences and derivatives, it is interesting to compare the Abel series with the general difference series studied in (vi). There, one is concerned with the coefficient functionals given by $\varDelta^n f(\beta n)$, for $-\infty < \beta < \infty$; the nature of the results obtained for the corresponding interpolation series depended in a complicated way upon the value of β. The analogy here would be to study the general Abel interpolation series associated with the sequence of values $f^{(n)}(\beta n)$. However, this time there is no gain in generality; except for the choice $\beta = 0$, which yields only the Maclaurin series, the substitution $g(z) = f(\beta z)$ reduces the general case immediately to the choice $\beta = 1$ which we have studied above.

It is possible to modify the Abel series in a useful way to apply to functions of type larger than 1. From Fig. 10, we observe that if f is an entire function of type less than π, then for sufficiently large $A > 0$, the function $g(z) = e^{Az} f(z)$ has its indicator diagram $D(g)$ within the set $\varOmega_\zeta$. If we then expand g into a Mittag-Leffler summable Abel series, we obtain

$$f(z) = e^{-Az}(\text{ML})\text{-}\sum_{n=0}^{\infty} \frac{z(z-n)^{n-1}}{n!}\, \mathscr{L}_n^*(f)$$

where

$$\mathscr{L}_n^*(f) = g^{(n)}(n) = (A\,e^A)^n \sum_{k=0}^n \binom{n}{k} A^{-k} f^{(k)}(n),$$

valid for any function of type less than π, or indeed, for any f obeying $h(\pm\pi/2) < \pi$.

(x) Laguerre polynomials. Instead of the generating relation used in § 9, we use[1]

$$(1-w)^{-1-\lambda}\exp\left(\frac{z\,w}{w-1}\right) = \sum_{n=0}^\infty L_n^{(\lambda)}(z)\,w^n. \qquad (10.8)$$

The parameter λ is usually taken to be real and greater than -1, but for our present purposes it can be any complex number, the $L_n^{(\lambda)}(z)$ then being defined by (10.8). We take the principal value of the power.

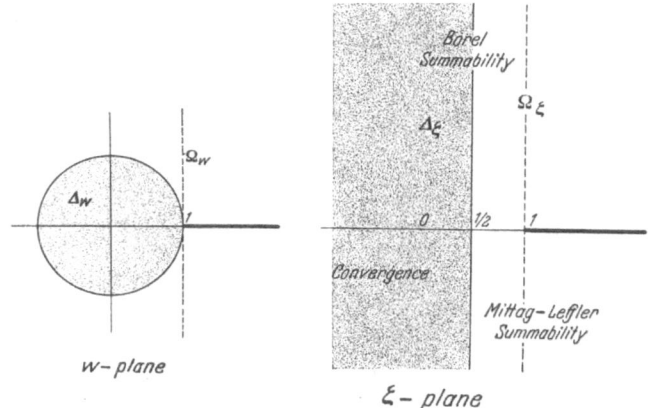

Fig. 11. Laguerre series

Here $g(w) = w/(w-1)$ is regular and univalent in the whole plane with the point $w=1$ deleted. When λ is not a negative integer, $A(w) = (1-w)^{-1-\lambda}$ has a singularity at $w=1$. We can take $\Omega_w = \Omega_w^*$ to be the w-plane, cut from 1 to $+\infty$ along the real axis; $\varDelta_w$ is $|w|<1$. Then $\Omega_\zeta = \Omega_\zeta^*$ is the ζ-plane, cut from 1 to $+\infty$, and $\varDelta_\zeta$ is the half-plane $\Re(\zeta) < \frac{1}{2}$. If λ is a negative integer, no cut is necessary, and Ω_ζ is the entire finite ζ-plane; but $\varDelta_\zeta$ is the same as before.

Hence the Laguerre expansion of an entire function of exponential type is convergent if $h(0) < \frac{1}{2}$; and Mittag-Leffler summable if $h(0) < 1$, or even if the conjugate indicator diagram avoids the ray $(1, +\infty)$. If λ is a negative integer, the Laguerre expansion of an entire function of exponential type is always Mittag-Leffler summable. Note, however,

---

[1] Erdelyi [1], vol. 3, pp. 250—251 (13); cf. also pp. 251—252 (21).

that what is often considered in the literature is the expansion in terms of the orthogonal functions $z^{\lambda/2} L_n^{(\lambda)}(z) e^{-z/2}$; cf. § 18.

It is interesting to compare these results with those obtained for the reversed Laguerre polynomials [§ 9, (iv)].

(xi) Angelescu polynomials[1]. These are generalizations of Laguerre polynomials, obtained by replacing the $A(w)$ of (10.8) by $\varphi\big(w/(1-w)\big)$. For $\varphi(s) = (1+s)^{1+\lambda}$ we have (10.8) again. Since $g(w)$ is the same as for the Laguerre polynomials, the expansion properties are the same except that zeros of $\varphi\big(w/(1-w)\big)$ lead to additional expansion formulas, while singularities of $\varphi\big(w/(1-w)\big)$ at points other than $w = 1$ introduce further restrictions on the class of functions that can be represented. We illustrate the possibilities by some further special cases.

(xii) Denisyuk polynomials[2]. (Singular point for A.) These are the Angelescu polynomials with $\varphi(s) = (s+1)/(2s+1)$, so that $A(w) = 1/(1+w)$. We can now take Ω_w to be the w-plane with the points $w = \pm 1$ removed; Ω_w^* is the w-plane with cuts from 1 to $+\infty$ and from -1 to $-\infty$ along the real axis; $\varDelta_w$ is $|w| < 1$; Ω_ζ is the ζ-plane with $\zeta = \frac{1}{2}$ removed; Ω_ζ^* is the ζ-plane cut from $\frac{1}{2}$ to $+\infty$; $\varDelta_\zeta$ is $\Re(\zeta) < \frac{1}{2}$. Since $\varDelta_\zeta$ is the same as for the Laguerre polynomials, the condition for convergence is the same as in (x); but the condition for Mittag-Leffler summability is now the more restrictive requirement that the conjugate indicator diagram avoids the ray $(\frac{1}{2}, +\infty)$. Another set considered by DENISYUK [5] has $A(w) = (1-w)/(1+w)^2$. This introduces no new expansion properties.

(xiii) Squared Hermite polynomials. If H_n are the usual Hermite polynomials [example (iii) of § 9], we have[3]

$$(1 - w^2)^{-\frac{1}{2}} \exp\left\{\frac{zw}{1-w}\right\} = \sum_{n=0}^{\infty} \{H_n[(-z/2)^{\frac{1}{2}}]\}^2 (-w/2)^n/n!.$$

Consequently $\{H_n[(-z/2)^{\frac{1}{2}}]\}$ are multiples of the Angelescu polynomials with $\varphi(s) = (s+1)(2s+1)^{-\frac{1}{2}}$. The expansion properties are the same as in example (xii).

(xiv) Adhoc polynomials. (A has a singularity closer to the origin than g does.) These are the Angelescu polynomials with $A(w) = (1-3w)/(1+2w)$. We take Ω_w to be the w-plane with $w = 1$ and $w = -\frac{1}{2}$ removed; then $\varDelta_w$ is $|w| < \frac{1}{2}$, Ω_ζ is the ζ-plane with $\zeta = \frac{1}{3}$ removed, and $\varDelta_\zeta$ is the disk with center at $\zeta = -\frac{1}{3}$ and radius $\frac{2}{3}$. Entire functions whose conjugate indicator diagrams lie in $|\zeta + \frac{1}{3}| < \frac{2}{3}$ are then represented

---

[1] ANGELESCU [1], SHASTRI [1], [2].

[2] DENISYUK [1] to [4].

[3] ERDÉLYI [1], vol. 3, p. 250 (12).

by convergent series of the $p_n(z)$, and indeed by more than one such series since $A(\tfrac{1}{3}) = 0$.

(xv) Actuarial polynomials[1]. (Multivalent g.) These are defined by $A(w) = e^{\lambda w}$, $g(w) = 1 - e^w$. We take Ω_w to be the strip $|v| < \pi$, so that Δ_w is $|w| < \pi$. Then $\Omega_\zeta = \Omega_\zeta^*$ is the ζ-plane cut from 1 to $+ \infty$ along the real axis, and Δ_ζ is the set where $|\log(1 - \zeta)| < \pi$. Since $|\log(1 - \zeta)| \leq -\log(1 - |\zeta|)$, Δ_ζ contains the disk $|\zeta| < 1 - e^{-\pi} = 0.957\ldots$. Thus a sufficient condition for convergence of the expansion of an entire function of exponential type τ is that $\tau < 1 - e^{-\pi}$, and a sufficient condition for Mittag-Leffler summability of the expansion is $h(0) < 1$. Because of the multivalence of g, there are an infinite number of expansions of zero, for example

$$0 = \sum_{n=1}^{\infty} p_n(z)\,(2k\pi i)^n, \qquad k = \pm 1, \pm 2, \ldots. \tag{10.9}$$

In the case $\lambda = 0$ these polynomials reduce effectively to the polynomials of TOUCHARD [1], generated by

$$e^{z(e^w - 1)} = \sum_{n=0}^{\infty} p_n(z)\,w^n. \tag{10.10}$$

(The numbers $n!\,p_n(1)$ with $p_n(z)$ from (10.10), are the so-called exponential numbers, (see also WILLIAMS [1], MIKSA [1]) and by taking $z = -1$ in (10.9) we have an infinite number of identities involving them.)

§ 11. More general polynomials

In all our examples so far, $\Psi(t)$ has been e^t. We now consider some examples with more general functions Ψ. Since $_0F_0(-; -; t) = e^t$, a more or less natural choice would be[2] the hypergeometric function $\Psi(t) = {}_pF_q(\alpha_1, \ldots, \alpha_p; \beta_1, \ldots, \beta_q; t)$ with $p \leq q + 1$. Here

$$_pF_q(\alpha_1, \ldots, \alpha_p; \beta_1, \ldots, \beta_q; t)$$

$$= {}_pF_q \begin{bmatrix} \alpha_1, \ldots, \alpha_p; \\ \beta_1, \ldots, \beta_q \end{bmatrix} t$$

$$= \sum_{n=1}^{\infty} \frac{(\alpha_1)_n \cdots (\alpha_p)_n}{(\beta_1)_n \cdots (\beta_q)_n} \frac{t^n}{n!},$$

where

$$(\gamma)_0 = 1, \quad (\gamma)_n = \gamma(\gamma+1)\ldots(\gamma+n-1) = \Gamma(\gamma+n)/\Gamma(\gamma).$$

With proper choice of $g(w)$, the polynomials $p_n(z)$ may also be hypergeometric functions.

---

[1] The name comes from TOSCANO [1]. Cf. ERDÉLYI [1], vol. 3, p. 254.
[2] For hypergeometric series see ERDÉLYI [1], vol. 1, chapter 4.

(xvi) **Special hypergeometric polynomials**[1]. Let us consider the polynomials defined by

$$(1 - w)^{-\lambda} \Psi \{- 4zw(1 - w)^{-2}\} = \sum p_n(z) w^n, \qquad (11.1)$$

where the power is to have its principal value. If $\lambda = 1$ and $\Psi(t) = {}_pF_q(a_1, \ldots, a_p; b_1, \ldots, b_q; t)$, then $p_n(z)$ is known to be a hypergeometric polynomial,

$$p_n(z) = {}_{p+2}F_{q+2}\left[\begin{array}{c} -n,\; n+1,\; a_1, \ldots, a_p; \\ \tfrac{1}{2},\quad .\quad 1,\quad b_1, \ldots, b_q; \end{array} z\right].$$

The special case $\Psi(t) = {}_1F_1(\tfrac{1}{2}; 1; t)$, $\lambda = 1$, yields *Bateman's polynomials*[2] $Z_n(z)$.

When $p \le q$, the Ψ of (11.1) is an entire function, and we have

$$A(w) = (1 - w)^{-\lambda}, \qquad g(w) = - 4w/(1 - w)^2.$$

Since $g(w)$ is regular and univalent, and $A(w)$ is regular, in $|w| < 1$, this disk will serve as both Ω_w and Δ_w. The image of $|w| < 1$ is the ζ-plane, cut from 1 to $+ \infty$. Hence we have the result that any entire function f whose Ψ-type is less than 1 can be expanded in a convergent series $\sum c_n p_n(z)$. For the particular case of the Bateman polynomials, $\Psi(t) = e^{t/2} J_0(it/2)$, which is of order 1 and type 1. In this case Ψ-type coincides with exponential type; if $f(z) = \sum_{n=0}^{\infty} a_n z^n/n!$,

$$F(\zeta) = \sum_{n=0}^{\infty} a_n \frac{(n!)^2 4^n}{(2n)! \zeta^{n+1}},$$

and the coefficients c_n in the expansion are defined by

$$c_n = \frac{(-1)^n}{\pi i} \int_\Gamma \{1 - (1 - \zeta)^{\frac{1}{2}}\}^{2n+1} \zeta^{-n-1} F(\zeta) \, d\zeta.$$

(xvii) **Reversed Bessel polynomials.** The Bessel polynomials[3] are generated by

$$(1 - 4zw)^{-\frac{1}{2}} \exp\left\{\frac{1 - (1 - 4zw)^{\frac{1}{2}}}{2z}\right\} = \sum_{n=0}^{\infty} p_n(z) w^n. \qquad (11.2)$$

They are a special case of the polynomials generated by the relation

$$(1 - 4zw)^{-\frac{1}{2}} \left\{\frac{2}{1 + (1 - 4zw)^{\frac{1}{2}}}\right\}^\alpha \Psi\left\{\frac{1 - (1 - 4zw)^{\frac{1}{2}}}{2z}\right\} = \sum_{n=0}^{\infty} p_n(z) w^n, \qquad (11.3)$$

---

[1] For $\lambda = 1$ see FASENMYER [1] and ERDÉLYI [1], vol. 1, pp. 193—194; vol. 3, p. 266.

[2] ERDÉLYI [1], vol. 1, pp. 193—194.

[3] ERDÉLYI [1], vol. 1, p. 195; vol. 3, p. 251; RAINVILLE [3]; OBRECHKOFF [1]. The name comes from KRALL and FRINK [1]; see also BURCHNALL [1].

which was discussed by RAINVILLE, chiefly in the case when Ψ is a hyper-geometric function, since in this case the p_n are multiples of hyper-geometric polynomials. We shall extend the usual terminology by calling any polynomials generated by (11.3) Bessel polynomials. They are not in generalized Appell form, but their reverses are, as we can see by the method of Lemma 9.6. For the reversed Bessel polynomials we have

$$A(w) = (1 - 4w)^{-\frac{1}{4}} \left\{ \frac{2}{1 + (1 - 4w)^{\frac{1}{2}}} \right\}^\alpha,$$

$$g(w) = \tfrac{1}{2} \{ 1 - (1 - 4w)^{\frac{1}{2}} \}.$$

For Ω_w we may take the w-plane cut from $\frac{1}{4}$ to $+\infty$, so that $\varDelta_w$ is $|w| < \frac{1}{4}$. The images Ω_ζ and $\varDelta_\zeta$ of these regions under $\zeta = g(w)$ are

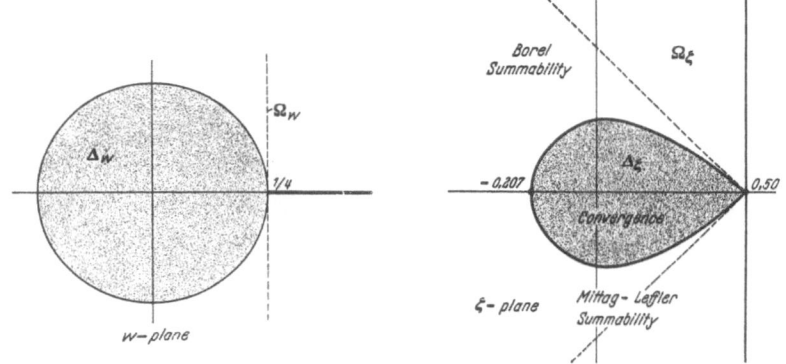

Fig. 12. w-plane for reversed Bessel polynomials   Fig. 13. ζ-plane for reversed Bessel polynomials

respectively the half-plane $\Re(\zeta) < \frac{1}{2}$ and the left-hand loop of the lemniscate $|\zeta - \zeta^2| < \frac{1}{4}$. The largest open disk lying in $\varDelta_\zeta$, with center at the origin, has radius $\frac{1}{2}(\sqrt{2} - 1) = 0.207\ldots$. Both Ω_w and Ω_ζ are star sets and hence coincide with Ω_w^* and Ω_ζ^*, respectively. Hence the expansion of an entire function of finite Ψ-type τ converges to the function if $\tau < \frac{1}{2}(\sqrt{2} - 1)$, and is Mittag-Leffler summable if $\tau < \frac{1}{2}$. These sufficient conditions can be weakened by requiring that the set $D(f)$ of § 2, with respect to the comparison function Ψ, lies in $\varDelta_\zeta$ or in Ω_ζ, respectively. For example, for the reverses of the original Bessel polynomials (11.2), the condition $h(0) < \frac{1}{2}$ is sufficient for Mittag-Leffler summability.

(xviii) **q-difference polynomials.** The polynomials satisfying the recursion

$$\frac{p_n(qz) - p_n(z)}{qz - z} = p_{n-1}(z) \qquad (11.4)$$

have been investigated by SHARMA and CHAK[1], who call them q-harmonic. They are Brenke polynomials generated by

$$A(w)\, e_q(z\,w) = \sum_{n=0}^{\infty} p_n(z)\, w^n,$$

$$e_q(t) = \sum_{n=0}^{\infty} \frac{w^n (q-1)^n}{(q^n-1)(q^{n-1}-1)\dots(q-1)}.$$

When $q \to 1$, (11.4) becomes the recursion relation $p_n'(z) = p_{n-1}(z)$ for Appell polynomials. When $0 < q < 1$, $e_q(t)$ is regular in $|t| < 1/(1-q)$, and the expansion theory of the polynomials belongs to the next chapter (§ 14). When $q > 1$, $e_q(t)$ is an entire function of order 0 and the polynomials come under our present theory. The functions of finite Ψ-type are a class of entire functions of order 0, in fact those functions $f(z) = \sum c_n z^n$ for which $c_n = O(q^{-n^2/2})$. How large a subclass of functions of finite e_q-type can be expanded depends on the location of the singular points of $A(w)$.

(xix) **Reversed Hermite polynomials.** The Hermite polynomials are defined by

$$\exp(2z\,w - w^2) = \sum_{n=0}^{\infty} H_n(z)\, w^n/n!.$$

Set $q_n(z) = H_n^*(z)/n!$, so that

$$e^{2w}\, e^{-z^2 w^2} = \sum_{n=0}^{\infty} q_n(z)\, w^n. \tag{11.5}$$

The $q_n(z)$ are not generalized Appell polynomials since $\Psi(t) = e^{-t^2}$ does not have non-zero coefficients. Correspondingly, the $q_n(z)$ do not form a basis, since they are all even and q_n and q_{n+1} are both of degree $2n$. If we set $p_n(z) = q_{2n}(z^{\frac12})$, (11.5) leads to

$$\cosh(2w^{\frac12})\, e^{-z\,w} = \sum p_n(z)\, w^n;$$

the $p_n(z)$ thus form an Appell set. Hence every entire function of exponential type can be expanded in a convergent series $\sum c_n p_n(z)$, and in infinitely many ways corresponding to the zeros of $\cosh(2w^{\frac12})$. When $f(z)$ is an even entire function, of growth at most order 2, finite type, $f(\sqrt{z})$ is an entire function of exponential type. Hence every even entire function of order 2 and finite type can be expanded in a convergent series of reversed Hermite polynomials of even index, and in infinitely many ways.

Since $H_{2n}(z) = (-1)^n 4^n n!\, L_n^{(-\frac12)}(x^2)$, the last statement is a special case of the result in (iv), § 9, about reversed Laguerre polynomials.

---

[1] SHARMA and CHAK [1]. The recursion (11.4) was suggested to us by T. S. MOTZKIN several years ago. See also BOAS and BUCK [1].

The reversed Hermite polynomials can be treated in another way. Set $Q_n(z) = q_n(iz) + zq_{n-1}(iz)$; then

$$e^{2w}(1 + zw)e^{z^2w^2} = \sum_{n=0}^{\infty} Q_n(z) w^n.$$

The Q_n are generalized Appell polynomials with $\Psi(t) = (1+t)e^{t^2}$, a function of order 2 and finite type. Hence every entire function of growth not exceeding order 2, finite type, has a convergent expansion of the form

$$\sum_{n=0}^{\infty} c_n z^n \{H_n(1/z) - iH_{n-1}(1/z)\}.$$

(xx) Rainville polynomials. These are the polynomials defined by

$$e^w I_0(zw) = \sum_{n=0}^{\infty} p_n(z) w^n,$$

or explicitly by

$$n! \, p_n(z) = (1 - z^2)^{n/2} P_n\{(1 - z^2)^{-\frac{1}{2}}\},$$

where P_n are Legendre polynomials. We discussed their reverses in (v), § 9. They are not generalized Appell polynomials, but

$$I_0\{(zw)^{\frac{1}{2}}\} \cosh(w^{\frac{1}{2}}) = \sum_{n=0}^{\infty} p_{2n}(z^{\frac{1}{2}}) w^n,$$

so that $p_{2n}(z^{\frac{1}{2}}) = q_n(z)$, say, are Brenke polynomials with $\Psi(t) = I_0(t^{\frac{1}{2}})$, $A(w) = \cosh(w^{\frac{1}{2}})$. Since $A(w)$ has an infinity of zeros, there are an infinite number of different representations of zero. Since $I_0(t^{\frac{1}{2}})$ is an entire function of order $\frac{1}{2}$, type 1, the functions of Ψ-type τ are those of growth not exceeding order $\frac{1}{2}$, type τ. For $\tau < 1$, such functions admit convergent expansions $\sum c_n q_n(z)$. Consequently all even entire functions of exponential type less than 1 have convergent expansions $\sum c_n p_{2n}(z) = \sum c_n(2n)! (1 - z^2)^n P_{2n}\{(1 - z^2)^{-\frac{1}{2}}\}$.

§ 12. Polynomials not in generalized Appell form

We have seen that some sets of polynomials that are not of generalized Appell form lead to related sets that are, and can therefore be discussed by our methods. Some other sets can be treated directly by modifications of our approach. This method seems to be most successful when there is some underlying periodicity either in the coefficient functionals $\mathscr{L}_n$ or in the polynomials p_n themselves. The first situation is illustrated by the Lidstone polynomials discussed in § 4. A simple example of the second is a set of polynomials studied by NASSIF [1]. Let $a_0, a_1, \ldots, a_{m-1}$ be given complex numbers, and set

$$p_{km+j}(z) = (z + a_j)^{km+j}, \quad j = 0, 1, \ldots, m - 1; \ k = 0, 1, \ldots.$$

To avoid computational complications we take up only the simple special case $m = 2$, $a_0 = 1$, $a_1 = -1$, so that $p_n(z) = (z + (-1^n)^n)$, a polynomial set introduced earlier by EWEIDA [1].

Set

$$H(z, w) = \sum_{n=0}^{\infty} p_n(z)\, w^n/n!$$

$$= \sum_{n=0}^{\infty} \frac{(z+1)^{2n}\, w^{2n}}{(2n)!} + \sum_{n=0}^{\infty} \frac{(z-1)^{2n+1}\, w^{2n+1}}{(2n+1)!}$$

$$= \cosh(z+1)\, w + \sinh(z-1)\, w$$

$$= e^{zw} \cosh w - e^{-zw} \sinh w .$$

Although the polynomials are not generalized Appell polynomials, we can write

$$\cosh(2w)\, e^{zw} = H(z, w)\cosh w + H(z, -w)\sinh w$$

$$= \sum_{n=0}^{\infty} p_n(z) \{\cosh w + (-1)^n \sinh w\}\, w^n/n! .$$

The process that we used for expanding a function in a series of generalized Appell polynomials still applies, and we obtain the following theorem, which goes considerably beyond the results of Nassif and Eweida.

Theorem 12.1. *Any entire function f of exponential type can be expanded in a convergent series of the form $\sum \mathcal{L}_n(f)\, p_n(z)$, where*

$$\mathcal{L}_n(f) = \frac{1}{2\pi i} \int_{\Gamma} \frac{w^n \{\cosh w + (-1)^n \sinh w\}}{n! \cosh(2w)}\, F(w)\, dw ,$$

F *is the Borel transform of f, and Γ encloses the conjugate indicator diagram of f and does not pass through a zero of $\cosh(2w)$.*

There are infinitely many different representations of zero, corresponding to the zeros of $\cosh(2w)$. The zeros nearest the origin are at $\pm \pi i/4$. Since the basic series is the expansion obtained when Γ encloses no zeros of $\cosh(2w)$, the basic series is applicable to entire functions of exponential type less than $\pi/4$, and this critical type cannot be replaced by any larger number. However, all that is really needed is to have the conjugate indicator diagram avoid all the zeros, so that it is enough, for example, to have $h(\pm \pi/2) < \pi/4$.

Chapter III

Representation of functions that are regular at the origin

§ 13. Integral representations

In Chapter II, the function Ψ used in defining a set of generalized Appell polynomials was itself an entire function; the functions for which we obtained expansions were in the class $\mathfrak{R}_\Psi$, which is a class of entire functions. In this chapter we take up the more complicated case in which Ψ is merely regular in some neighborhood of the origin. We suppose again that $\Psi(t) = \sum_{n=0}^{\infty} \Psi_n t^n$ with $\Psi_n > 0$, we suppose that $\lim \Psi_n^{1/n}$ exists (finite), and we suppose, by way of normalization, that thie limit is 1. Then $\Psi(t)$ is regular for all t in a set E that contains the open disk $|t| < 1$ and has $t = 1$ as a boundary point. In Chapter II ws singled out for study the class $\Re_{\Psi}$ of functions $f(z) = \sum f_n z^n$ of finite Ψ-type, i.e. such that $\limsup |f_n/\Psi_n|^{1/n} < \infty$. The notion of finite Ψ-type is no longer of much use: the relation $\limsup |f_n/\Psi_n|^{1/n} = R$ now merely asserts that f is analytic in the disk $|z| < 1/R$, regardless of the choice of Ψ, so that $\Re_{\Psi}$ is the set of *all* functions f that are regular at the origin. We find it more useful to study the class $\mathfrak{A}(\Omega)$ consisting of all functions which are analytic in a specified simply-connected neighborhood Ω of the origin. Certain fundamental differences are at once apparent between the present situation and that studied before. In Chapter II, an expansion that represented a given entire function represented it everywhere. This will not always be true now: an expansion formula that applies to every member of a class $\mathfrak{A}(\Omega)$ may represent some members of the class only in a proper subset of Ω. For example, the Maclaurin series applies to every member of $\mathfrak{A}(\Omega)$, but represents it only in the largest open disk, centered at 0, which is contained in Ω. We take this as a pattern, and associate with an expansion formula defined by polynomials $\{p_n\}$ and coefficient functionals $\{\mathscr{L}_n\}$, and applying to all elements of $\mathfrak{A}(\Omega)$, a subregion Ω_0 such that any function f that is regular in Ω has the representation $\sum \mathscr{L}_n(f) p_n(z)$, uniformly convergent on compact subsets of Ω_0. In contrast, WHITTAKER and his students have dealt almost exclusively with the case in which Ω is a disk and Ω_0 coincides with Ω [1]. (An expansion with this property is said to be effective in Ω.) Our approach yields more precise results for some sets of polynomials whose natural expansion regions are not circular.

Let $f(z) = \sum_{n=0}^{\infty} f_n z^n$, regular at the origin. We define F, the generalized Borel transform of f, by

$$F(w) = \sum_{n=0}^{\infty} \frac{f_n}{\Psi_n w^{n+1}}. \tag{13.1}$$

If f is regular in $|z| < r$ then $\limsup |f_n/\Psi_n|^{1/n} \le 1/r$ and $F(w)$ is regular at least for $|w| > 1/r$. Let $V(f)$ be the set of points λw_0 with $0 \le \lambda \le 1$ and w_0 a singularity of F, and let $\Omega^{\bullet}$ be the smallest closed set which contains all the sets $V(f)$ for f in $\mathfrak{A}(\Omega)$. If f is regular in Ω, then F is regular at least outside $\Omega^{\bullet}$. If Ω contains the open disk $|z| < r$, then

---
[1] But see FALGAS [1].

$\Omega^\bullet$ is a closed subset of the disk $|w| \leq 1/r$; and if w_0 is in $\Omega^\bullet$ so is λw_0 for $0 \leq \lambda \leq 1$; that is, $\Omega^\bullet$ is star-shaped with respect to the origin.

We denote the complement of the set S by S'. We use $S_1 \cdot S_2$ to mean the set of all points of the form $z_1 z_2$ with $z_1 \in S_1$, $z_2 \in S_2$, and $1/S$ to mean the set of all points z^{-1} such that $z \in S$; finally, S_1/S_2 means $S_1 \cdot (1/S_2)$. Using this notation, we can state the following form of the Hadamard multiplication theorem[1].

Theorem 13.2. *Let Λ_1 and Λ_2 be simply connected neighborhoods of the origin and let f and g be regular in Λ_1 and Λ_2, respectively, with $f(z) = \sum f_n z^n$ and $g(z) = \sum g_n z^n$. Then $h(z) = \sum f_n g_n z^n$ has a regular extension to any simply connected neighborhood of the origin lying in the set $(\Lambda_1' \cdot \Lambda_2')'$.*

An immediate consequence of this is the following fundamental result connecting functions in $\mathfrak{A}(\Omega)$ with their generalized Borel transforms.

Theorem 13.3. *Let Ψ be regular in the region Λ, let $\Omega^\oplus = (\Lambda'/\Omega^\bullet)'$ and let C be any compact subset of $\Omega^\oplus$. Then there exists a contour Γ surrounding $\Omega^\bullet$ such that for any f regular in Ω and any z in C,*

$$f(z) = (2\pi i)^{-1} \int_\Gamma \Psi(zw) F(w)\, dw.$$

The set $\Omega^\oplus$ is a subset of Ω and often coincides with it; it may alternatively be described as the set of all z in Ω such that $z \cdot \Omega^\bullet \subset \Lambda$. In some cases $\Omega^\oplus$ and $\Omega^\bullet$ are independent of the choice of Ψ.

Theorem 13.4. *If Ω is the disk $|z| < r$, then, for all choices of Ψ such that $\lim \Psi_n^{1/n} = 1$, the set $\Omega^\bullet$ is the disk $|w| \leq 1/r$ and $\Omega^\oplus = \Omega$.*

For, let us consider the particular function $f(z) = \sum\limits_{n=0}^{\infty} (z/r)^{n!}$. This is regular in Ω and its transform $F(w) = \sum\limits_{n=0}^{\infty} (\Psi_n r^{n!} w^{n!+1})^{-1}$ has $|w| = 1/r$ as a natural boundary, independently of the choice of Ψ. Thus $V(f)$ is the disk $|w| \leq 1/r$. Since no $V(f)$ can be smaller than this and $\Omega^\bullet$ is the intersection of all $V(f)$, $\Omega^\bullet$ is the same disk. To determine $\Omega^\oplus$, we use the fact that z is in $\Omega^\oplus$ if and only if $zw \in \Lambda$ for all w with $|w| \leq 1/r$. Since 1 is a boundary point of Λ, $\Omega^\oplus$ is also the disk $|w| \leq 1/r$.

For general regions Ω we must take the nature of Ψ into account (cf. Theorem 13.5). One special choice of Ψ is of considerable importance. In Theorem 10.2 we showed that the kernel $K(z, w) = H\{a(w) + zg(w)\}$ generates generalized Appell polynomials if and only if $H(t) = e^t$ or $H(t) = (1-t)^{-\lambda}$. The first possibility [with $g(w) = w$] led to the Pólya representation for entire functions of exponential type and to Appell polynomials. We shall now consider the second possibility. It leads to a class of generalized Appell polynomials that seem to be "natural" polynomials for expanding functions which are analytic at the origin,

[1] See DIENES [1], p. 346; BIEBERBACH [1], p. 300; BIEBERBACH [2].

in the same way that Appell polynomials seem to be natural polynomials for expanding entire functions of exponential type. [Another natural connection between the respective kernel functions is that $e^t = {}_0F_0(-;-;t)$ and $(1-t)^{-\lambda} = {}_1F_0(\lambda;-;t)$.]

For $\Psi(t) = (1-t)^{-\lambda}$, and indeed for a somewhat more general class of functions Ψ, we can obtain a simple characterization of the sets $\Omega^\bullet$ and $\Omega^\oplus$.

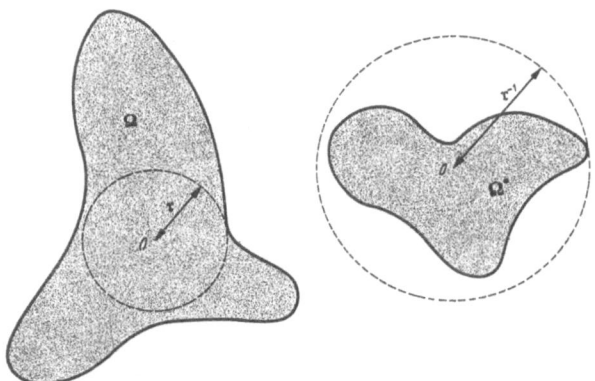

Fig. 14. Special case $\Omega^\bullet = 1/\Omega'$

Theorem 13.5. *If Ω is star-shaped with respect to the origin, $\Psi(t) = \sum \Psi_n t^n$, $\Psi(t)$ is regular except for $1 < t < \infty$, and $1/\Psi_n$ are the Hausdorff moments of a function of bounded variation,*

$$1/\Psi_n = \int_0^1 u^n d\mu(u), \qquad \int_0^1 |d\mu(u)| < \infty, \tag{13.6}$$

then $\Omega^\bullet = 1/\Omega'$ and $\Omega^\oplus = \Omega$.

When Ω is not (as it was in Theorem 13.4) a disk $|w| < r$, but contains such a disk, then for the class of functions Ψ just described, the set $\Omega^\bullet$ is a proper subset of the closed disk $|w| \le 1/r$ (see Fig. 15). For example, if Ω is the open disk $|w-1| < 2$, $\Omega^\bullet$ is the closed disk $|w + \frac{1}{3}| \le \frac{2}{3}$.

Both the conditions imposed on Ψ in Theorem 13.5 are satisfied for $\Psi(t) = (1-t)^{-\lambda}$ if $\lambda \ge 1$, since

$$(-1)^n \Big/ \binom{-\lambda}{n} = \frac{n!\,(\lambda-1)!}{(\lambda+n-1)!} = (\lambda-1)\int_0^1 u^n (1-u)^{\lambda-2} du, \qquad \lambda > 1,$$

and the case $\lambda = 1$ is trivial. Both conditions are also satisfied when

$$1/\Psi_n = \frac{(\alpha+\beta)!}{\alpha!\,(\beta-1)!} \int_0^1 (1-u)^{\beta-1} u^{\alpha+n} du = \frac{(\alpha+\beta)!\,(\alpha+n)!}{\alpha!\,(\alpha+\beta+n)!}$$

if $\alpha > -1$ and $\beta > 0$; in this case $\Psi(t)$ is a hypergeometric function (reducing to $(1-t)^{-\lambda}$ for $\alpha = 0$, $\beta = \lambda - 1$).

Since $1/\Psi_n$ are Hausdorff moments, we have, if $f(z) = \sum\limits_{n=0}^{\infty} f_n z^n$,

$$w F(w) = \sum_{n=0}^{\infty} (f_n/\Psi_n)\, w^{-n} = \sum_{n=0}^{\infty} f_n w^{-n} \int_0^1 u^n \, d\mu(u)$$

$$= \int_0^1 \left\{ \sum_{n=0}^{\infty} f_n (u/w)^n \right\} d\mu(u),$$

$$w F(w) = \int_0^1 f(u/w) \, d\mu(u). \tag{13.7}$$

From (13.7) we see that if f is regular in Ω, then F is regular at least for all w with $u/w \in \Omega$ for $0 \leq u \leq 1$. Since Ω is star-shaped, $uz \in \Omega$ for $0 \leq u \leq 1$ if $z \in \Omega$; therefore F is regular for all w such that $1/w \in \Omega$. This shows that $V(f) < 1/\Omega'$ and $\Omega^\bullet < 1/\Omega'$.

Since Λ is the plane cut from 1 to ∞, $\Lambda'/\Omega^\bullet < \Lambda' \cdot \Omega' = \Omega'$, and $\Omega^\oplus = (\Lambda'/\Omega^\bullet)' > \Omega$. Since $\Omega^\oplus \subset \Omega$, this means that $\Omega^\oplus = \Omega$. Therefore $\Lambda'/\Omega^\bullet = 1/\Omega^\bullet = \Omega'$, and so $\Omega^\bullet = 1/\Omega'$.

By specializing Theorem 13.3 we now obtain an analogue of the Pólya representation.

Theorem 13.8. *Let Ω be star-shaped with respect to the origin and let C be a compact subset of Ω. If Ψ has the properties assumed in Theorem 13.5, and in particular if*

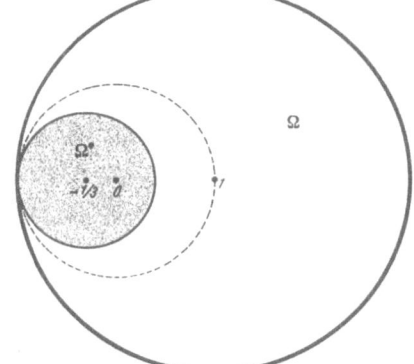

$\Psi(t) = (1-t)^{-\lambda}$, $\lambda > 1$, *there is a contour Γ surrounding the set $1/\Omega'$ such that for $z \in C$ every function that is regular in Ω admits the representation*

$$f(z) = (2\pi i)^{-1} \int_\Gamma \Psi(zw) F(w) \, dw,$$

with F regular in $1/\Omega$ [and defined by (13.7)].

§ 14. Brenke polynomials

Brenke polynomials (cf. § 6) are the generalized Appell polynomials for which $g(w) = w$, so that they are generated by

$$A(w)\, \Psi(zw) = \sum_{n=0}^{\infty} p_n(z)\, w^n. \tag{14.1}$$

They have a particularly transparent theory since no complications are introduced by the special properties of $g(w)$.

We shall use (14.1) in the form

$$\Psi(zw) = \sum_{n=0}^{\infty} p_n(z) \, w^n/A(w). \qquad (14.2)$$

In the examples discussed in Chapter II, the set of points (z, w) for which this series converges is a cylindrical set $\varXi$ in 4-space, whose base is a set $\varOmega$ in the w-plane and whose axis is in the z-direction. When Ψ is merely regular at the origin, the series does not converge for all z for a given w, and the convergence set $\varXi$ may be more difficult to describe.

For example, the trivial expansion $\Psi(zw) = (1 - zw)^{-1} = \sum_{n=0}^{\infty} z^n w^n$ converges for all (z, w) with $|zw| < 1$.

Let $\varOmega$ be star-shaped with respect to the origin and let (14.2) be convergent for $(z, w) \in \varXi$. We determine the set $\varOmega^{\bullet}$ and then consider the set $\varOmega_0$ consisting of all points $z \in \varOmega$ such that $(z, w) \in \varXi$ for all $w \in \varOmega^{\bullet}$. If $(z, w) \in \varXi$ then $zw \in \varLambda$, so that $\varOmega_0 \subset \varOmega^{\oplus}$. If we take the integral representation of Theorem 13.3 and replace $\Psi(zw)$ by the series (14.2) we obtain the following analogue of Theorem 7.2.

Theorem 14.3. *If Ψ has the expansion* (14.2), *any function f in $\mathfrak{A}(\varOmega)$ has the expansion*

$$f(z) = \sum \mathscr{L}_n(f) \, p_n(z), \qquad (14.4)$$

convergent in $\varOmega_0$ and uniformly convergent in compact subsets, with

$$\mathscr{L}_n(f) = (2\pi i)^{-1} \int_{\varGamma} w^n F(w) \, \{A(w)\}^{-1} \, dw, \qquad (14.5)$$

where $F(w)$ is defined by (13.1) *and $\varGamma$ surrounds $\varOmega^{\bullet}$.*

We may discuss summability by replacing the set $\varXi$ by the set $\varXi^*$ of all points (z, w) for which (14.2) is true with summability instead of convergence, uniformly in compact subsets.

Zeros of $A(w)$ in $\varOmega^{\bullet}$ give rise to representations of zero, valid for $z \in \varOmega_0$ (instead of for all z as in Chapter II).

We illustrate Theorem 14.3 with some special classes of Brenke polynomials.

(i) **Polynomials generated by** $A(w)\,(1-zw)^{-\lambda}$. The Brenke polynomials with $\Psi(t) = (1-t)^{-\lambda}$ are particularly simple, and have many analogies with ordinary Appell polynomials[1]. We first note some of their formal properties.

Theorem 14.6. *Let $p_n(z)$ be the polynomials defined by the formal power-series relation*

$$A(w)\,(1 - zw)^{-\lambda} = \sum_{n=0}^{\infty} p_n(z)\, w^n, \qquad \lambda > 0, \qquad (14.7)$$

$$A(w) = \sum_{n=0}^{\infty} a_n w^n, \qquad a_0 \neq 0.$$

---

[1] Cf. Boas and Buck [1].

Then

$$p'_n(z) = \lambda p_{n-1}(z) + z p'_{n-1}(z), \tag{14.8}$$

and $p_n(z)$ is the reversed $(C, \lambda - 1)$ partial sum[1] of $A(w)$. In particular, when $\lambda = 1$,

$$p_n(z) = \sum_{k=0}^{n} a_k z^{n-k} \tag{14.9}$$

and

$$\frac{p_n(z) - p_n(0)}{z} = p_{n-1}(z), \quad n \geq 1. \tag{14.10}$$

The recursion formula (14.8) is obtained immediately by differentiating (14.7) with respect to z and comparing coefficients. The explicit form of $p_n(z)$ is of course a special case of (6.4), but can be obtained more directly as follows:

$$
\begin{aligned}
A(w)(1 - zw)^{-\lambda} &= \sum_{n=0}^{\infty} a_n w^n \sum_{k=0}^{\infty} \binom{k + \lambda - 1}{\lambda - 1} z^k w^k \\
&= \sum_{j=0}^{\infty} w^j \sum_{s=0}^{j} a_{j-s} \binom{s + \lambda - 1}{\lambda - 1} z^s \\
&= \sum_{j=0}^{\infty} w^j p_j(z);
\end{aligned}
$$

we have the desired result on comparing the coefficients of w^j in the last two series.

Another connection between the $p_n(z)$ and Appell polynomials is shown by the polynomials $q_n(z) = p_n(z/\lambda)$, which satisfy

$$q'_n(z) = q_{n-1}(z) + (z/\lambda) q'_{n-1}(z),$$

so that $\lim_{\lambda \to \infty} q_n$ is the Appell polynomial of degree n corresponding to $A(w)$.

The relation (14.10) is the special case $q = 0$ of the recursion

$$\frac{p_n(qz) - p_n(z)}{qz - z} = p_{n-1}(z), \tag{14.11}$$

which we discussed for $q > 1$ in § 11. The case $0 < q < 1$ is considered below (example ii).

We now consider expansions in series of the polynomials defined by (14.7) with $\lambda \geq 1$, using Theorem 14.3 and the description of $\Omega^\bullet$ given in Theorem 13.5. [In the degenerate case $A(w) = 1$, $\lambda = 1$, $p_n(z) = z^n$ and our expansion is simply the Maclaurin series.] If $A(w)$ is regular in a region Σ_w, the expansion (14.7) converges for all points (z, w) with $|w| < \varrho_0$ and $|zw| < 1$, where ϱ_0 is the distance from the origin to the boundary of Σ_w. Let Ω be star-shaped with respect to the origin. The set Ω_0 consists of all $z \in \Omega$ such that $|zw| < 1$ for all $w \in \Omega^\bullet = 1/\Omega'$ (by

---

[1] See, e.g., HARDY [1], pp. 96—97.

Theorem 13.5). Let r_0 be the distance from the origin to the boundary of Ω and suppose that $r_0\varrho_0>1$. Then Ω_0 is the disk $|z|<r_0$. Hence we have the following theorem.

   Theorem 14.12. *If polynomials $p_n(z)$ are defined by (14.7) with $\lambda\geq 1$, if $A(w)$ is regular in the disk $|w|<\varrho_0$, and if Ω contains the disk $|z|\leq 1/\varrho_0$, then every function f regular in Ω can be expanded in a series of the $p_n(z)$, convergent in the largest open disk having its center at the origin and lying in Ω.*

   If $A(w)$ has zeros in $|w|<\varrho_0$ there will be nontrivial representations of zero valid for $|z|<1/\varrho_0$; in this case a function that can be expanded in accordance with Theorem 14.12 can be so expanded in more than one way, at least in $|z|<1/\varrho_0$.

   When $A(w)$ is entire, there is no restriction on Ω and the situation is the same as for the Maclaurin series. When $A(w)$ is not entire, a restriction on Ω is necessary, as the following example shows. Let

$$(1 - 2w)^{-1}(1 - zw)^{-1} = \sum p_n(z)\,w^n,$$

so that $p_n(z) = (z^{n+1} - 2^{n+1})/(z-2)$. Since $A(w)$ has a singular point at $w = \frac{1}{2}$, Theorem 14.2 asserts that any function which is regular on the closed disk $|z|\leq 2$ can be expanded in a series of the $p_n(z)$. However, this is not true for a smaller disk. For, if a series $\sum b_n\,p_n(z)$ converges at $z=0$, then the series $\sum 2^n\,b_n$ converges, and the explicit form of the $p_n(z)$ shows that $\sum b_n\,p_n(z)$ will in fact converge in $|z|<2$. Hence no function that is not at least regular in $|z|<2$ could have a convergent expansion $\sum b_n\,p_n(z)$.

   (ii) q-difference polynomials. We now discuss the polynomials satisfying (14.11) for $0<q<1$. They are generated by

$$A(w)\,e_q(zw) = \sum_{n=0}^{\infty} p_n(z)\,w^n, \tag{14.13}$$

$$e_q(t) = E_q\big(t(1-q)\big) = \sum_{n=0}^{\infty} \frac{\{t(1-q)\}^n}{(1-q^n)(1-q^{n-1})\dots(1-q)}.$$

The function $E_q(t)$ is a degenerate case $(a=0)$ of the "basic" hypergeometric series[1] $_1\Phi_0(a, z)$, and has the representation

$$E_q(t) = \prod_{n=0}^{\infty} (1 - q^n t)^{-1},$$

which shows that it is regular in the plane cut along the positive real axis from 1 to ∞. Thus $E_q(t)$ satisfies one of the hypotheses of Theorem

---

[1] ERDÉLYI [1], vol. 1, pp. 195—196.

13.5. It satisfies the other hypothesis since

$$\sigma_n = 1/\Psi_n = (1 - q^n)(1 - q^{n-1}) \ldots (1 - q)$$

is a completely monotonic sequence[1].

To prove this, we use the difference operator ∇ defined by $\nabla f(n) = f(n) - f(n+1)$. We have

$$\nabla\{q^n f(n)\} = q^n\{1 - q + q\nabla\}f(n) \tag{14.14}$$

and hence

$$\nabla^r\{q^n f(n)\} = q^n\{1 - q + q\nabla\}^r f(n).$$

To say that σ_n is completely monotonic is to say that $\nabla^r \sigma_n \geq 0$ for all nonnegative integers r, n. Now $\sigma_n \geq 0$, and

$$\nabla\sigma_n = \sigma_n - (1 - q^{n+1})\sigma_n = q^{n+1}\sigma_n \geq 0.$$

Now, by applying (14.14), we obtain

$$\nabla^2\sigma_n = q\nabla(q^n\sigma_n) = q^{n+1}\{1 - q + q\nabla\}\sigma_n,$$

and generally,

$$\nabla^{r+1}\sigma_n = q^{n+1}\{1 - q + q\nabla\}^r\sigma_n.$$

Since $0 < q < 1$, we obtain from this that $\nabla^{r+1}\sigma_n \geq 0$ if $\nabla^j\sigma_n \geq 0$ for $j = 0, 1, \ldots, r$; the desired conclusion follows by induction[2].

Now

$$\begin{aligned} A\left(w/(1-q)\right)E_q(zw) &= A\left(w/(1-q)\right)E_q\left(w(1-q)\cdot z/(1-q)\right) \\ &= A\left(w/(1-q)\right)e_q\left(w\cdot z/(1-q)\right), \end{aligned}$$

so the polynomials generated by (14.13) can be more conveniently thought of as generated by

$$A\left(w/(1-q)\right)E_q(zw) = \sum_{n=0}^{\infty} w^n p_n(z)/(1-q)^n.$$

Since $E_q(t)$ satisfies the hypotheses of Theorem 13.5, the expansion properties of $p_n(z)$ are exactly the same as in Theorem 14.12, except that we now must require that Ω contains the disk $|z| < \{\varrho_0(1-q)\}^{-1}$, where $A(w)$ is regular in $|w| < \varrho_0$.

§ 15. More general polynomials

We now consider generalized Appell polynomials without any specializations other than those imposed in § 13. Thus

$$A(w)\Psi(zg(w)) = \sum_{n=0}^{\infty} p_n(z)w^n, \tag{15.1}$$

---

[1] For the representation of a completely monotonic sequence as a sequence of Hausdorff moments see e.g. WIDDER [1], p. 108.

[2] N. J. FINE has shown us an explicit construction of a nondecreasing $\alpha(t)$ such that $\sigma_n = \int_0^1 t^n d\alpha(t)$.

where A and g are regular at the origin, $A(0) \neq 0$, $g(0) = 0$, $g'(0) = 1$; and $\Psi(t)$ is regular in a set Λ, with $\Psi(t) = \sum_{n=0}^{\infty} \Psi_n t^n$ in a neighborhood of the origin, $\Psi_n > 0$, $\lim \Psi_n^{1/n} = 1$. Let Ω_w be a region in which $A(w)$ and $g(w)$ are both regular, and in which $g(w)$ is univalent. Let Ω_ζ be the image of Ω_w under the map $\zeta = g(w)$. For fixed z, the left-hand side of (15.1) is regular, as a function of w, for all $w \in \Omega_w$ such that $zg(w) \in \Lambda$. Let $\Delta_w(\varrho)$ be an open disk $|w| < \varrho$ contained in Ω_w and let $\Delta_\zeta(\varrho)$ be its image in Ω_ζ. Then the series in (15.1) converges as a power series in w for $|w| < \varrho$ whenever z is such that $z\Delta_\zeta(\varrho) < \Lambda$. The set of all such points z will be denoted by $C(\varrho)$; we observe that $C(\varrho) = \left(\Lambda'/\Delta_\zeta(\varrho)\right)'$. Let $w = W(\zeta)$ be the inverse of $\zeta = g(w)$ and set $B(\zeta) = A\left(W(\zeta)\right)$. We then have the following statement, in which $\times$ denotes the Cartesian product.

Theorem 15.2. *The series*

$$B(\zeta)\, \Psi(z\zeta) = \sum p_n(z)\, \{W(\zeta)\}^n \tag{15.3}$$

converges for all points $(z, \zeta) \in C(\varrho) \times \Delta_\zeta(\varrho)$ *for any* ϱ *such that* $|w| < \varrho$ *is in* Ω_w.

We write (15.3) in the form

$$\Psi(z\zeta) = \sum p_n(z)\, \{W(\zeta)\}^n / B(\zeta) \equiv \sum p_n(z)\, u_n(\zeta), \tag{15.4}$$

generalizing (14.2). We denote by Ξ the set of points (z, ζ) for which the series converges, and repeat the considerations leading up to Theorem 14.3. The set Ω_0 is now the set of all points $z \in \Omega$ such that $(z, \zeta) \in \Xi$ for all $\zeta \in \Omega^{\bullet}$. Theorem 14.3 is replaced by the following result.

Theorem 15.5. *If* Ψ *has the expansion* (15.4)*, and* $\Omega < \Omega_w$*, any function* f *in* $\mathfrak{A}(\Omega)$ *has the expansion*

$$f(z) = \sum \mathscr{L}_n(f)\, p_n(z), \tag{15.6}$$

convergent in Ω_0 *and uniformly convergent in compact subsets, with*

$$\mathscr{L}_n(f) = (2\pi i)^{-1} \int_\Gamma u_n(\zeta)\, F(\zeta)\, d\zeta, \tag{15.7}$$

where $u_n(\zeta)$ *are defined in* (15.4) *and* $F(\zeta)$ *is defined by* (13.1) *(with* ζ *in place of* w*) and* Γ *surrounds* $\Omega^{\bullet}$.

In applying Theorems 15.2 and 15.5 to a specific set of polynomials, we proceed as follows. We start with the generating relation (15.1) and a region Ω. We determine the set $\Omega^{\bullet}$ by using properties of Ψ and Ω as in § 13. The functions Ψ and g together determine the sets $C(\varrho)$ and $\Delta_\zeta(\varrho)$ and thus the convergence set Ξ described after Theorem 15.2. From Ξ and $\Omega^{\bullet}$ we find the set Ω_0 in which the expansion formula (15.6) converges for any $f \in \mathfrak{A}(\Omega)$. Finally, the zeros of A correspond to distinct independent representations of zero. The zeros of $B(\zeta)$ in Ω_ζ are isolated. so that the countour Γ of Theorem 15.5 can be supposed to avoid them. The coefficients $\mathscr{L}_n(f)$ will depend on the zeros of $B(\zeta)$ that are inside Γ,

§ 16. Polynomials generated by $A(w)\left(1-zg(w)\right)^{-\lambda}$

We consider the polynomials defined by

$$A(w)\left(1-zg(w)\right)^{-\lambda} = \sum p_n(z)\, w^n. \qquad (16.1)$$

Let Ω be star-shaped with respect to 0. We can use Theorem 13.5 to determine $\Omega^\bullet$. With the notations of § 15, $C(\varrho)$ is the set $1/\{\Delta_\zeta(\varrho)\}'$, Ξ is the union of the sets $1/\{\Delta_\zeta(\varrho)\}' \times \Delta_\zeta(\varrho)$, and $z\in\Omega_0$ if $(z,\zeta)\in\Xi$ for all $\zeta\in\Omega^\bullet = 1/\Omega'$. The sharpest results are obtained if Ω is taken to be one of the sets $C(\varrho)$; then $\Omega^\bullet$ will be $\Delta_\zeta(\varrho)$, so that $\Omega_0 = \Omega$. We can then state the following special case of Theorem 15.5.

Theorem 16.2. *Let polynomials* $p_n(z)$ *be defined by* (16.1), *with* $A(0)\neq 0$, $g(0)=0$, $g'(0)\neq 0$. *Let* Ω_w *be a region in which* A *and* g *are regular and* g *is univalent; let* $G(\varrho)$ *be the region containing the origin and bounded by the image of* $|w|=\varrho$ *under the mapping* $z=1/g(w)$, *with* $|w|\leq\varrho$ *contained in* Ω_w *and* $G(\varrho)$ *star-shaped with respect to the origin. Then any function regular in* $G(\varrho)$ *can be represented in* $G(\varrho)$ *as a series of the polynomials* $p_n(z)$.

We illustrate Theorem 16.2 with some examples.

(iii) **Taylor series.** We consider the trivial case[1] $p_n(z)=(z-c)^n$. The generating relation is

$$\sum_{n=0}^{\infty} (z-c)^n\, w^n = \left(1-(z-c)\,w\right)^{-1} = (1+c\,w)^{-1}\left(1-zg(w)\right)^{-1},$$

with $g(w)=w/(1+cw)$. To compute $G(\varrho)$, we write $z=1/g(w)=(1+cw)/w=c+w^{-1}$; the image of $|w|=\varrho$ is the circle $|z-c|=1/\varrho$. Since Ω_w is the plane with $w=-1/c$ deleted, ϱ must be chosen so that $\varrho<1/|c|$; then the disk $|z-c|<1/\varrho$ contains the origin, and any function regular there can be expanded in a Taylor series about c. The result fails to be completely sharp because our discussion has been restricted to expansions of functions that are regular at the origin.

(iv) **Lerch polynomials**[2]. The polynomials generated by

$$\{1-z\log(1+w)\}^{-\lambda} = \sum_{n=0}^{\infty} p_n(z)\, w^n$$

are expressible in terms of the Laplace transforms of Newton polynomials. Here $A(w)=1$ and $g(w)=\log(1+w)$. For Ω_w we can take the w-plane, cut from -1 to ∞ along the negative real axis. The image

---

[1] This has also been discussed by BOURBAKI [1].
[2] ERDÉLYI [1], vol. 3, p. 258 (4), (5).

of $|w| = \varrho$ under $z = 1/\log(1+w)$ is the set where $|e^{1/z} - 1| = \varrho$; we require $\varrho < 1$. Then the expansion of any function regular in a region $|e^{1/z} - 1| < \varrho < 1$ is convergent in that region. In particular, since the region is inside the disk $|z| < 1/\log(1+\varrho)$, any function analytic in this disk has an expansion which converges at least in the corresponding region (not necessarily in the whole disk).

(v) **Gegenbauer polynomials**[1]. These may be defined by

$$(1 - 2zw + w^2)^{-\lambda} = \sum_{n=0}^{\infty} C_n^{(\lambda)}(z)\, w^n,$$

which for our purposes it is preferable to write in the form

$$(1 + w^2)^{-\lambda}\left\{1 - z\,\frac{2w}{1+w^2}\right\}^{-\lambda} = \sum C_n^{(\lambda)}(z)\, w^n.$$

We may take Ω_w to be $|w| < 1$, and then $G(\varrho)$ is the complement of the set of points $z = (1+w^2)/(2w)$ for which $|w| \leq \varrho$. Computation shows that this is the elliptical region $\{2\varrho/(1+\varrho^2)\}^2 x^2 + \{2\varrho/(1-\varrho^2)\}^2 y^2 < 1$ with foci at ± 1. We thus obtain the known result[2] that any function analytic in the interior of an ellipse with foci at ± 1 can be represented there by a convergent series of Gegenbauer polynomials. Legendre polynomials are the special case $\lambda = \frac{1}{2}$.

We do not obtain a stronger result here by replacing convergence by Mittag-Leffler summability since $\Omega_w = \Omega_w^*$.

(vi) **Chebyshev polynomials**[3]. These may be defined by the generating relation

$$-\tfrac{1}{2}\log(1 - 2zw + w^2) = \sum_{n=1}^{\infty} n^{-1} T_n(z)\, w^n,$$

which is not of generalized Appell form. However, if we differentiate this with respect to w, we obtain

$$1 + 2\sum_{n=1}^{\infty} T_n(z)\, w^n = \frac{1 - w^2}{1 + w^2}\left\{1 - z\,\frac{2w}{1+w^2}\right\}^{-1},$$

which is of the type we are considering in this section. The same discussion as for the Gegenbauer polynomials applies and the conclusion is identical.

(vii) **Humbert polynomials**[4]. These are defined by

$$(1 - 3zw + w^3)^{-\lambda} = \sum_{n=1}^{\infty} p_n(z)\, w^n,$$

[1] ERDÉLYI [1], vol. 2, pp. 174ff.; vol. 3, pp. 246—247.
[2] Special case of SZEGÖ [1], p. 238, Theorem 9.1.1.
[3] ERDÉLYI [1], vol. 2, pp. 183ff., especially p. 186 (30).
[4] ERDÉLYI [1], vol. 3, p. 246. For $\lambda = \frac{1}{2}$ the polynomials were considered by PINCHERLE.

which we write in the form

$$(1 + w^3)^{-\lambda}\left(1 - z\,\frac{3\,w}{1 + w^3}\right)^{-\lambda} = \sum_{n=1}^{\infty} p_n(z)\,w^n. \tag{16.3}$$

Here, although $A(w)$ and $g(w)$ are regular in $|w| < 1$, $g(w)$ is not univalent there. We may take Ω_w to be $|w| < 1/\sqrt[3]{2}$, and then $G(\varrho)$ is the interior of the curve $3z = w^{-1} + w^2$ with $|w| = \varrho$. This is a hypotrochoid (for $\varrho = 1/\sqrt[3]{2}$ it is a three-cusped hypocycloid), with parametric equations

$$\left.\begin{aligned} 3\,x &= \varrho^{-1}\cos\theta + \varrho^2\cos 2\theta, \\ 3\,y &= -\varrho^{-1}\sin\theta + \varrho^2\sin 2\theta. \end{aligned}\right\} \tag{16.4}$$

Thus any function analytic in the interior of one of these curves (with $\varrho < 1/\sqrt[3]{2}$) can be represented there by a convergent series of Humbert polynomials. The curve (16.4) is inside the circle with center at 0 and radius $\frac{1}{3}(1 + \varrho^3)/\varrho$, and surrounds the circle with radius $\frac{1}{3}(1 - \varrho^3)/\varrho$. Thus in particular if f is regular in a disk $|z| < R$ with $R > \sqrt[3]{\frac{1}{4}}$, then its Humbert expansion converges to f at least in the disk $|z| < r$ where $r < R$ and $(R - r)(R + r)^2 = \frac{8}{27}$. For large R, $r \sim R - \frac{2}{27}R^{-2}$.

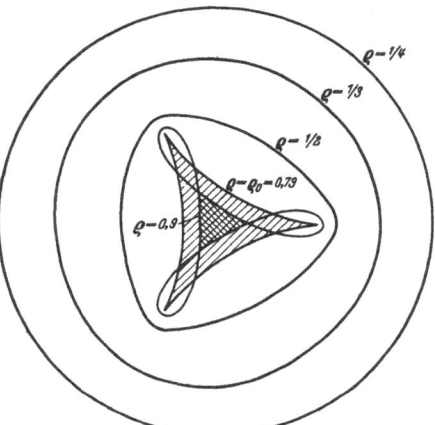

Fig. 16. Hypotrochoids

The nonunivalence of $g(w)$ introduces a phenomenon analogous to that observed in §8. Let w_1 and w_2 be two points in $|w| < 1$ at which $g(w_1) = g(w_2)$; then provided that (16.3) converges at z for both w_1 and w_2, we obtain a nontrivial representation of zero, and the values of z for which this occurs fill a region. In fact, for $1 > \varrho > 1/\sqrt[3]{2}$ the series (16.3) converges, if $|w| < \varrho$, for z in the region, containing the origin, bounded by part of the curve (16.4) (which is now a hypotrochoid with loops). Hence for z in such a region there are an infinity of representations of zero and thus an infinity of different expansions of any function which is analytic in a sufficiently large region.

(viii) **Faber polynomials**[1]. Consider a closed, bounded, simply-connected convex region Ω with complement Ω'. Let $t = f(z) = z + f_0 + f_1 z^{-1} + f_2 z^{-2} + \cdots$ map Ω' on the circular region $|t| > \varrho$, and let $z = h(t)$

---

[1] FABER [1], SCHIFFER [1], BOURION [1], FALGAS [3].

be the inverse function. The Faber polynomials $F_n(z)$ associated with Ω [or with $f(z)$] can be defined by

$$\frac{h'(t)}{h(t) - z} = \sum_{n=0}^{\infty} F_n(z) \, t^{-n-1}.$$

If we put $w = 1/t$, this is

$$\frac{h'(1/w)}{w \, h(1/w)} \left\{ 1 - \frac{z}{h(1/w)} \right\}^{-1} = \sum_{n=0}^{\infty} F_n(z) \, w^n,$$

so that Faber polynomials are of the form (16.1) with $\lambda = 1$, $A(w) = w^{-1} h'(1/w)/h(1/w)$, and $g(w) = 1/h(1/w)$. The choice $h(w) = (w^2 + 1)/(2w)$ leads to the Chebyshev polynomials again.

Since $A(w)$ is regular where $g(w)$ is, in applying Theorem 16.2 we can take $\Omega_w = \{|w| < 1/\varrho\}$. The mapping $z = 1/g(w)$ is $z = h(1/w)$ and the image of $|w| = \tau$ under this is the level line $|f(z)| = 1/\tau$, $\tau > \varrho$. A function which is regular in Ω is regular in the interior of some level line $|f(z)| = 1/\tau$, i.e. in some $G(\tau)$, and can therefore be represented there by a series of the Faber polynomials $F_n(z)$. We thus have, for convex regions, a proof of FABER's expansion theorem for Faber polynomials; it is not essentially different from FABER's proof, but illustrates how the theorem fits into our general framework. Summability of Faber expansions could be discussed in the same way[1].

§ 17. Special hypergeometric polynomials

We consider again the hypergeometric polynomials obtained by taking Ψ to be a generalized hypergeometric function ${}_pF_q$ of argument $-4zw/(1-w)^2$. In § 11 we considered the case $p \leq q$, when ${}_pF_q$ was an entire function. Now we consider the case $p = q + 1$, when ${}_pF_q(t)$ is regular except for $1 \leq t < \infty$. Setting $z = 1/g(w) = (1-w)^2/(-4w)$, we see that the images of $|w| = \varrho$, for $\varrho < 1$, are the ellipses with foci at 0 and 1. Hence by generalizing the reasoning of Theorem 14.12, we see that the polynomials (11.1), with $p = q + 1$, serve to represent any function analytic in such an ellipse provided that the parameters of the ${}_pF_q$ satisfy conditions (which we do not attempt to specify in the general case) making $\{1/\Psi_n\}$ a sequence of Hausdorff moments.

(ix) **Jacobi polynomials.** In particular, we have

$$(1 - w)^{-1-\alpha-\beta} {}_2F_1 \left[\begin{matrix} \dfrac{1+\alpha+\beta}{2}, & \dfrac{2+\alpha+\beta}{2} \\ 1+\alpha \end{matrix} \; ; \; \dfrac{-4wz}{(1-w)^2} \right]$$
$$= \sum_{n=0}^{\infty} \frac{(\alpha+\beta+n)! \, \alpha!}{(\alpha+\beta)! \, (\alpha+n)!} \, P_n^{(\alpha,\beta)}(1 - 2z) \, w^n,$$

---

[1] Cf. also SEWELL [1], LOHIN [1], ALPER [1]. The uniqueness of Faber expansions is discussed by SUYETIN [1].

where $P_n^{(\alpha,\beta)}$ are the Jacobi polynomials[1]. Since the coefficients Ψ_n satisfy the hypotheses of Theorem 13.5 if $\beta > 0$ and $\alpha > -1$ (cf. pp. 50–51), we see that any function analytic in an ellipse with foci at 0 and 1 is representable as a series in $P_n^{(\alpha,\beta)}(1-2z)$; in other words, any function analytic in an ellipse with foci at 1 and -1 is representable as a series in $P_n^{(\alpha,\beta)}(z)$ if $\alpha > -1$, $\beta > 0$. Since $P_n^{(\beta,\alpha)}(z) = (-1)^n P_n^{(\alpha,\beta)}(z)$, this holds also if $\alpha > 0$, $\beta > -1$. To obtain the complete known result[2] that this holds if $\alpha > -1$, $\beta > -1$, we appeal to the fact that

$$2 \frac{d}{dz} P_n^{(\alpha-1,\beta-1)}(x) = (n + \alpha + \beta - 1) P_{n-1}^{(\alpha,\beta)}(x),$$

so that by expanding $f'(z)$ in terms of $P_n^{(\alpha,\beta)}(z)$ we can deduce an expansion of $f(z)$ in terms of $P_n^{(\alpha-1,\beta-1)}(z)$. The restriction $\alpha > -1$, $\beta > -1$ is now seen to be irrelevant for the expansion of analytic functions; we are not using the orthogonality of the polynomials, which is all that is lost for smaller values of α and β.

§ 18. Polynomials not in generalized Appell form

Just as for expansions of entire functions, the general theory serves to elucidate the behavior of some sets of polynomials which are not in the form of generalized Appell polynomials. As an instance of this possibility, consider the set given by $p_0 = 1$, $p_n(z) = 1 + z^n$ $(n > 0)$; this is used as an example by WHITTAKER[3]. The function

$$K(z, w) = \sum_{n=0}^{\infty} p_n(z) w^n = 1 + \sum_{n=1}^{\infty} (1 + z^n) w^n$$
$$= (1 - zw)^{-1} + w/(1 - w)$$

has singularities at $w = 1$ and $w = 1/z$. If $|z| < r$, then K is regular for $|w| < 1/r$ except at $w = 1$. Thus the series expansion for $K(z, w)$ converges for all $|w| < 1/r$ when $r > 1$, but only for $|w| < 1$ when $r \leq 1$ (as one sees also by direct inspection of the series). If f is regular in Ω then F is regular outside $\Omega^{\bullet} = 1/\Omega'$. In the integral representation of Theorem 13.7, Γ must lie outside $\Omega^{\bullet}$; it must also be an admissible path, on which the series for $K(z, w)$ converges. In the present case these requirements are irreconcilable unless $r > 1$. Thus if f is regular in the disk $|z| < r$ with $r > 1$, it can be expressed in the form $f(z) = \sum c_n p_n(z)$.

A more general situation involving polynomials that are not in generalized Appell form arises as follows[4]. Consider the Appell poly-

---

[1] ERDÉLYI [1], vol. 2, pp. 168ff.; vol. 3, p. 264 (9). The $P_n^{(\alpha,\beta)}(z)$ themselves cannot be put in generalized Appell form unless $\alpha - \beta = 0$, 1, or -1. (SMITH [1].)

[2] SZEGÖ [1], p. 238.

[3] WHITTAKER [2], p. 12.

[4] Cf. BOAS [2] for a different, but equivalent, approach.

nomials generated by

$$A(w)\, e^{zw} = \sum_{n=0}^{\infty} p_n(z)\, w^n.$$

If we write this as

$$e^{zw} = \sum_{n=0}^{\infty} p_n(z)\, u_n(w), \qquad u_n(w) = w^n / A(w), \tag{18.1}$$

and then apply a linear transformation U to both sides, qua functions of w, we obtain a new generating relation of the form

$$\Psi(z, w) = \sum_{n=0}^{\infty} p_n(z)\, U_n(w). \tag{18.2}$$

If the transformation is appropriately chosen, (18.2) may allow us to expand functions which (18.1) did not. In particular, if we can get $\Psi(z, w) = 1/(w - z)$, (18.2) will clearly be suitable for expanding functions that are regular at 0, whereas (18.1) is, as we have seen, suitable primarily for entire functions of exponential type. We can achieve the desired result formally by taking the Laplace transform of (18.1), but this will succeed only if $1/A(w)$ is sufficiently small at ∞.

We have

$$\int_0^{\infty} e^{zw} e^{-ws}\, dw = 1/(s - z), \tag{18.3}$$

provided $\Re(s - z) > 0$, where the path of integration is understood to be the positive real axis. If we rotate the line of integration so that $w = r e^{i\alpha}$, $0 \le r < \infty$, the condition for convergence is that $\Re\{(s - z)e^{i\alpha}\} > 0$, in other words that $s - z$ is in the right-hand half plane rotated through an angle α in the negative direction.

Suppose now that (18.1) converges uniformly after multiplication by e^{-ws}, when $w = r e^{i\alpha}$, for all z in a set Ω, and for all s such that $\Re\{(s - z)e^{i\alpha}\} > 0$. Then when $\Re\{(s - z)e^{i\alpha}\} > 0$ we have

$$(s - z)^{-1} = \sum_{n=0}^{\infty} p_n(z) \int_0^{\infty e^{i\alpha}} u_n(w)\, e^{-ws}\, dw = \sum_{n=0}^{\infty} p_n(z)\, U_n^{(\alpha)}(s). \tag{18.4}$$

If Γ is a contour in the s-plane that can be divided into arcs Γ_k on each of which (18.4) holds for some $\alpha = \alpha_k$, and $z \in \Omega$, then with $U_n(s) = U_n^{(\alpha_k)}(s)$ on Γ_k we have

$$(s - z)^{-1} = \sum_{n=0}^{\infty} p_n(z)\, U_n(s), \qquad z \in \Omega, \tag{18.5}$$

on Γ, and Γ is an admissible path for this expansion (for $z \in \Omega$).

If now $f(z)$ can be represented for $z \in \Omega$ by its Cauchy integral along Γ, we obtain an expansion for $f(z)$ in a series of the $p_n(z)$ by using (18.5) and integrating term by term.

A simple sufficient condition that will make the transition from (18.1) to (18.5) possible is that

$$\sum_{n=0}^{\infty} |p_n(z)| \frac{|w|^n}{|A(w)|} e^{-\Re(ws)} \tag{18.6}$$

converges uniformly for z in Ω, s on each arc Γ_k, and $w = r e^{i\alpha_k}$. In verifying (18.6) we have to have enough information about the asymptotic behavior of $p_n(z)$ and about the magnitude of $1/A(w)$. For a simple example, take the Appell polynomials with $A(w) = (1-w)^\lambda$, considered in § 5. Here we consider only the case where λ is a positive integer. From the integral formula for the coefficients of a power series we have

$$|p_n(z)| \leq (1+R)^\lambda e^{R|z|} R^{-n}$$

with arbitrary positive R, and the series (18.6) is dominated, for $|w - 1| > \varepsilon > 0$, by

$$\sum_{n=0}^{\infty} (1+R)^\lambda e^{R|z|} R^{-n} |w|^n e^{-\Re(ws)}.$$

If we take $R = |w| + \delta$, $\delta > 0$, we need to have, for z in Ω, $\Re(rse^{i\alpha_k}) > r|z|$ for all $r > 0$ and s on Γ_k. Let Ω be a disk $|z| < h$; then we need $\Re(se^{i\alpha_k}) > h$ for s on Γ_k [and of course $\alpha_k \neq 0$ so that the line of integration in (18.4) will avoid the singular point of $1/A(w)$]. We can make the segments Γ_k collectively approximate the circumference $|s| = h$ from outside, as closely as we please. Consequently every function analytic in $|z| \leq h'$, $h' > h$, can be represented in $|z| < h$ by a series of the polynomials $p_n(z)$. In other words, the set $\{p_n(z)\}$ represents every function analytic at 0 in the same region as its Maclaurin series. This accounts for the facts that we observed in § 1 about the polynomials (1.5).

When $A(w)$ is a more general entire function we have to avoid points where $A(w)$ is small by taking a curvilinear path of integration in (18.4). The simplest case is that in which $A(w)$ is an entire function of order less than 1. In this case it is well known that $|A(w)| \geq e^{-\varepsilon(w)}$, where $\varepsilon(w) = o(1)$, outside a set of circles whose total length is finite. Let $f(z)$ be regular in Ω, a disk $|z| < h$; we again approximate the circumference $|s| = h$ in the s-plane from outside by a convex polygon Γ made up of line segments Γ_k, choosing α_k so that (18.4) holds. In (18.4) we are going to take the paths of integration to be the rays $\theta = \alpha_k$ modified by indentations of finite total length so that the paths avoid the exceptional circles for $A(w)$. We now need to show that (18.6) converges uniformly for z in Ω, s on each Γ_k and w on each of the indented rays $\theta = \alpha_k$.

We need an asymptotic estimate for $p_n(z)$. This is easily obtained. We have

$$A(w) e^{zw} = \sum_{n=0}^{\infty} p_n(z) w^n. \tag{18.7}$$

If A is an entire function of exponential type, with indicator $h(\varphi)$, the left-hand side of (18.7) is (for each z) an entire function of exponential type whose indicator is $h(\varphi) + r\cos(\theta + \varphi)$, where $z = re^{i\theta}$; consequently its type is

$$\max_{\varphi} |h(\varphi) + r\cos(\theta + \varphi)| = k(z), \qquad (18.8)$$

say. On the other hand the type can be computed from the coefficients in the series on the right-hand side of (18.7), and is $\limsup |n!\, p_n(z)|^{1/n}$. Thus we have

$$\limsup |n!\, p_n(z)|^{1/n} = k(z), \qquad (18.9)$$

where $k(z)$ is given by (18.8)[1].

In the case at hand, when A is of order less than 1, $h(\varphi) = 0$ and $k(z) = |z|$. Then (18.6) is dominated, for large $|w|$, by

$$\sum_{n=0}^{\infty} e^{-\Re(ws)}\, \frac{(|z| + \delta)^n}{n!}\, |w|^n\, e^{\varepsilon(w)},$$

where δ is positive but arbitrarily small, s is on Γ_k and w is on the indented ray $\theta = \alpha_k$. This dominant series converges uniformly provided that $\Re(ws) > h$ for these values of s and w, and this holds for large $|w|$.

Hence if Ω is the disk $|z| < h$ we can use (18.5) to infer a representation of $f(z)$ in a series of the $p_n(z)$, converging in any interior disk. In other words, a function that is regular at 0 has an expansion in terms of the $p_n(z)$, which converges in the same disk as its Maclaurin expansion[2].

The same conclusion holds if $A(w)$ is an entire function of zero exponential type[2], the only difference being that we have to use a more complicated estimate for $1/A(w)$. When $A(w)$ is an entire function of order 1 and mean type, the behavior of $1/A(w)$ is less simple, and all that can be inferred is that[2] if $f(z)$ is regular in a sufficiently large neighborhood of 0 its expansion in terms of the $p_n(z)$ converges in some neighborhood of 0. If $A(w)$ is no longer of exponential type, the estimate (18.9) for $p_n(z)$ breaks down, but if $A(t)$ has enough directions of slow decrease the proof can still be carried through[3]. For example, if $A(t) = e^{t^3}$, every function regular in a triangle[4] with vertices at $-k$, $k\omega$, $-k\omega^2$, where $k > 0$ and ω is an imaginary cube root of -1, is represented by its expansion at least inside the inscribed circle of this triangle.

---

[1] SHEFFER [2].

[2] SHEFFER [2], BOAS [2].

[3] BOAS [2].

[4] An attempt (BOAS [2]) to apply similar reasoning to Sheffer polynomials is fallacious.

Although the general expansion theory of Hermite and Laguerre polynomials has not been discussed by our methods, we summarize the results here for reference[1].

A necessary and sufficient condition for the analytic function $f(z)$ to be represented by a convergent series of the Hermite polynomials [§ 9, (iii)] in the strip $|y| < \tau$ is that to each β, $0 \leq \beta < \tau$, there corresponds a finite positive $B(\beta)$ such that

$$|f(x + iy)| \leq B(\beta) \exp\left\{\frac{y^2 - x^2}{2} - |x| (\beta^2 - y^2)^{\frac{1}{2}}\right\}.$$

If a series of Hermite polynomials converges for any non-real points, it converges in a strip.

For Laguerre polynomials, the natural representation region is a parabola; it is simpler to state the conditions for $f(z^2)$ to be represented in a strip by a series $\sum c_n L_n(z^2)$: *a necessary and sufficient condition for $f(z^2)$ to be represented by a convergent series of Laguerre polynomials $L_n(z^2) = L_n^{(0)}(z^2)$ in the strip $|y| < \tau$ is that to each β, $0 \leq \beta < \tau$, there corresponds a finite positive $B(\beta)$ such that*

$$|f(z^2)| \leq B(\beta) \exp\{\tfrac{1}{2} x^2 - |x| (\beta^2 - y^2)^{\frac{1}{2}}\}.$$

If a series $\sum c_n L_n(z^2)$ converges for any non-real points, it converges in a strip.

Chapter IV

Applications

We conclude by giving two examples of classes of problems that can be solved by using expansions in series of Appell polynomials or generalized Appell polynomials.

§ 19. Uniqueness theorems

Our first example illustrates the advantages of considering summable expansions. The theorems in question state that an entire function of sufficiently slow growth must vanish identically if some sequence of linear functionals vanish when applied to it. The oldest theorem of this kind is CARLSON's theorem, which states that an entire function of exponential type, such that $\limsup |y|^{-1} \log |f(it)| < \pi$, vanishes identically if it vanishes at the nonnegative integers[2]. A proof which lends itself readily to generalization by the methods of this survey is

---

[1] HILLE [1] for Hermite polynomials, POLLARD [1] for Laguerre polynomials.

[2] See BOAS [3], Chapter 9.

as follows. According to § 10 (vi), such a function $f(z)$ has for all z a Mittag-Leffler summable expansion in Newton polynomials,

$$f(z) = \sum_{n=0}^{\infty} \binom{z}{n} \Delta^n f(0);$$

since all $\Delta^n f(0)$ are zero if all $f(n) = 0$, the hypotheses of CARLSON'S theorem imply that $f(z) \equiv 0$.

(DEMAR has recently shown that uniqueness theorems of this type can also be obtained directly from the integral representations of the functionals, using a modification of the argument used in the proof of Theorem 8.8. By avoiding the use of expansions, one obtains more general results. See DEMAR [4].)

Similarly, if $\{p_n(z)\}$ is a set of generalized Appell polynomials, and $f(z)$ has a Mittag-Leffler summable expansion with coefficients $\mathscr{L}_n(f)$, we have immediately that $f(z) \equiv 0$ if all $\mathscr{L}_n(f) = 0$. For example[1], $f(z) \equiv 0$ if $f^{(n)}(n) = 0$ for all nonnegative integers n, provided that the conjugate indicator diagram of f is inside the region Ω_ϱ of § 10 (ix) associated with Mittag-Leffler summability of Abel interpolation series, and in particular if f is of exponential type less than 1.

Another example of somewhat similar nature was given at the end of § 4, where we saw that an entire function of finite exponential type which obeys $f^{(2n)}(0) = f^{(2n)}(1) = 0$ for $n = 0, 1, 2, \ldots$ must be a finite sum $\sum C_k \sin(k\pi z)$.

The Bernoulli series, introduced in § 9, (i), can also be used in this fashion. There, we showed that any entire function of finite type has a convergent representation in the form

$$f(z) = \sum_{n=0}^{\infty} q_n(z) \mathscr{L}_n(f) + \sum_{k=1}^{N} C_k e^{2\pi k z i},$$

where

$$\mathscr{L}_n(f) = \int_0^1 f^{(n)}(t)\, dt.$$

Consequently, if $\mathscr{L}_n(f) = 0$ for $n = 0, 1, 2, \ldots$, then f is a finite exponential sum. This condition is satisfied if f is periodic, with period 1; the converse also holds, for if $\mathscr{L}_n(f) = 0$ for $n = 1, 2, \ldots$, then $g^{(n)}(0) = 0$ for $n = 0, 1, \ldots$ where $g(z) = f(z+1) - f(z)$. Accordingly, this provides yet another approach to the well known characterization of the periodic entire functions of exponential type.

The operators $\mathscr{L}_n$ for which uniqueness theorems can be obtained in this way are those which appear as coefficients in terms of sets of

---

[1] GELFOND; see BUCK [3].

generalized Appell polynomials as in (7.4). In particular, for Appell polynomials they are of the form

$$\mathscr{L}_n(f) = \frac{1}{2\pi i} \int_\Gamma \frac{w^n}{A(w)} F(w)\, dw,$$

or formally

$$\mathscr{L}_n(f) = \frac{1}{A(D)} f^{(n)}(z)\Big|_{z=0}$$

(cf. § 20). When there are nontrivial representations of zero (§ 8), we also have $f(z) \equiv 0$ if the values of $\mathscr{L}_n(f)$ are the coefficients in a representation of 0.

§ 20. Functional equations

Let $A(w) = \sum_{n=0}^\infty a_n w^n$, $a_0 \neq 0$, let D stand for d/dz, and consider the differential equation (in general, of infinite order)

$$A(D) y = f(z), \tag{20.1}$$

where $f(z)$ is regular in a region Ω and we seek a solution y which is regular in a subset of Ω; the precise sense in which $A(D)$ is to be interpreted will be specified later. Formally we have

$$y = \frac{1}{A(D)} f(z), \tag{20.2}$$

and we want to show that under appropriate hypotheses the right-hand side of (20.2) can be given a meaning, that $A(D)y$ then exists in an appropriate sense, and that (20.1) holds. It turns out that the expansion theory of Appell polynomials associated with $A(w)$ is a convenient tool for this discussion[1].

We consider first the particularly simple case when f is an entire function of exponential type and $A(w)$ is a polynomial, so that (20.1) is an ordinary linear differential equation with constant coefficients. Let $p_n(z)$ be the Appell polynomials generated by $A(w)$. Let $f(z)$ be an entire function of exponential type. Then

$$f(z) = \sum_{n=0}^\infty c_n p_n(z) \tag{20.3}$$

where the series is convergent, uniformly on compact sets (Theorem 9.2). Now we have

---

[1] SHEFFER [1] to [4].

$$p_n(z) = \frac{1}{2\pi i} \int_\gamma \frac{A(w)\, e^{zw}}{w^{n+1}}\, dw,$$

where γ is a contour surrounding the origin. This suggests that $\frac{1}{A(D)}\, p_n(z) = z^n/n!$, so we are led to consider the function

$$y(z) = \sum_{n=0}^{\infty} c_n z^n/n!. \tag{20.4}$$

Now we know that in (20.3) we can take

$$c_n = \frac{1}{2\pi i} \int_\Gamma \frac{w^n}{A(w)}\, F(w)\, dw,$$

where Γ is a contour on which $A(w) \neq 0$, and F is the Borel transform of f. Since $c_n = O(R^n)$ for some R, the series in (20.4) converges uniformly on compact sets and defines an entire function $y(z)$ of exponential type. The explicit formula of § 6 for Appell polynomials can be written

$$p_n(z) = A(D)\, (z^n/n!),$$

so from (20.4) we have at once

$$A(D)\, y = \sum_{n=0}^{\infty} c_n p_n(z) = f(z).$$

We thus have a solution of (20.1) corresponding to every Appell expansion (20.3) of $f(z)$. If $A(w)$ is not constant, it has zeros, and each zero gives rise to an independent representation of zero as a series $\sum c_n p_n(z)$, and hence to an independent solution of (20.1).

The discussion can be extended to the case where $A(w)$ is merely regular at 0; the series (20.3) is then Mittag-Leffler summable, uniformly on compact sets, if the conjugate indicator diagram of f is inside the Mittag-Leffler star of $A(w)$ (this is the specialization of Theorem 7.7 corresponding to Theorem 9.2). Before going into details we need to define the symbol $A(D)\, y$ when y is an entire function of exponential type and A is not necessarily a polynomial. Let y have the Pólya representation

$$y(z) = \frac{1}{2\pi i} \int_\Gamma e^{zw}\, Y(w)\, dw;$$

then it is natural to define

$$A(D)\, y = \frac{1}{2\pi i} \int_\Gamma A(w)\, e^{zw}\, Y(w)\, dw, \tag{20.5}$$

since the right-hand side has the formal expansion

$$\sum_{n=0}^{\infty} a_n y^{(n)}(z).$$ (20.6)

If $\varGamma$ is inside the circle of convergence of the Maclaurin series of A, (20.6) is convergent; and if $\varGamma$ is inside the Mittag-Leffler star of A, (20.6) is Mittag-Leffler summable; we may take (20.6), instead of (20.5), as the definition of $A(D) y$.

Now if $\varGamma$ is a contour in the Mittag-Leffler star of A on which $A(w)$ is different from 0, and which surrounds the conjugate indicator diagram of f, we have $F(w)$ regular on $\varGamma$, and

$$\sum_{n=0}^{\infty} \frac{(zw)^n}{n!} \frac{F(w)}{A(w)} = e^{zw} F(w)/A(w)$$

converges uniformly for z in any compact set and w on $\varGamma$. Hence

$$\frac{1}{2\pi i} \int_{\varGamma} e^{zw} \frac{F(w)}{A(w)} dw = \sum_{n=0}^{\infty} \frac{z^n}{n! \, 2\pi i} \int_{\varGamma} \frac{w^n}{A(w)} F(w) \, dw = \sum_{n=0}^{\infty} c_n z^n/n!. \quad (20.7)$$

The right-hand side is the series (20.4). Thus $y(z)$ is defined by (20.4) as an entire function of exponential type, and (20.7) provides, in accordance with (20.5), a justification of (20.2). Furthermore, again in accordance with (20.5), we have $A(D) y = f(z)$, by using the left-hand side of (20.7).

It is interesting to note that if $y(z)$ is given to begin with, there is a series definition of $A(D) y$ more general than (20.6): write (20.5) in the form

$$A(D) \, y = \frac{1}{2\pi i} \int_{\varGamma} A(w) \, e^{(z-t) \, w} \, e^{t \, w} \, Y(w) \, dw;$$

then if $\varGamma$ is in the Mittag-Leffler star of A and surrounds the conjugate indicator diagram of y, we can write

$$A(w) \, e^{(z-t) \, w} = (\text{ML}) - \sum_{n=0}^{\infty} w^n \, p_n(z-t)$$

(Mittag-Leffler summability) and deduce

$$A(D) \, y = (\text{ML}) - \sum_{n=0}^{\infty} p_n(z-t) \, y^{(n)}(t).$$

For $t = z$ we have (20.6); for $t = 0$ we have an expansion of $A(D) y$ in terms of the polynomials $p_n(z)$:

$$A(D) \, y = \sum_{n=0}^{\infty} y^{(n)}(0) \, p_n(z).$$

For example, if $A(w) = (1 + e^w)^{-\lambda}$, the polynomials are generalized Eulerian polynomials [§ 9 (i)] and[1] $A(D)$ is formally $(2\nabla)^{-\lambda}$, where $\nabla f = \frac{1}{2}[f(z+1) + f(z)]$.

The same idea can be used in other situations[2]. Consider for example, the difference equation

$$\sum_{j=1}^{k} a_j\, y(z + \omega_j) = f(z), \tag{20.8}$$

which is of the form (20.1) with

$$A(w) = \sum_{j=1}^{k} a_j\, e^{\omega_j w},$$

where $f(z)$ is not necessarily an entire function. Let $\varrho(z) = \max|z + \omega_j|$, and let z^* be the point where $\varrho(z^*)$ is smallest. If f is regular at z^* it can be shown that the expansion (20.3) of $f(z)$ in terms of Appell polynomials converges in a neighborhood of z^*, and using this expansion we arrive at a solution of (20.8) in a neighborhood of z^*. If $f(z)$ is regular at some other point $z = c$, we find a solution regular at $c - z^*$, valid in a circle of radius exceeding $|z^*|$.

If this approach is combined with a uniqueness theorem, results of some delicacy can be easily obtained. As an illustration, consider the following very simple difference equation

$$y(z + 1) - \beta\, y(z) = f(z). \tag{20.9}$$

Here $\beta\ (\neq 0)$ is a constant, and f is an entire function of finite exponential type obeying $h(\pm \pi/2; f) < \pi$. We seek a solution $y(z)$ which satisfies the same condition. It is clear that if we specify $y(0) = A$, then (20.9) enables us to compute all of the values $\{y(n)\}$, $n = 0, 1, 2, \ldots$. Since conditions are appropriate for an appeal to CARLSON's theorem discussed in the preceding section, we can immediately conclude that if there is a solution $y(z)$ of the desired sort, it is necessarily unique. However it is not evident (or true) that there *is* such a solution. The complete story is as follows: *when β is any real or complex number, except a negative real, then there is a solution $y(z)$ of (20.9) with arbitrary assigned initial value $y(0) = A$ and satisfying $h(\pm \pi/2; y) < \pi$; when β is real and negative, then there is not a solution of (20.9) with this growth restriction unless $\sum_0^{\infty} f(k)\beta^{-k} = 0$; when this condition on β is satisfied, then there is exactly one solution of (20.9) obeying the required growth restriction.*

---

[1] SUMNER [1], BOAS [4].
[2] See SHEFFER [4].

[Note that in this last case, $y(0)$ is determined.] The details of the proof of this may be found in BUCK [1].

Bibliography

ALPER, S. YA: [1] On uniform approximation of functions of a complex variable on a closed region. Izv. Akad. Nauk SSSR., Ser. Mat. **19**, 423—444 (1955) [Russian].

AL-SALAM, W. A.: [1] The Bessel polynomials. Duke Math. J. **42**, 529—545 (1957). (§11, xvii.)

ANGELESCU, A.: [1] Sur certains polynomes généralisant les polynomes de Laguerre. C. R. Acad. Sci. Roum. **2**, 199—201 (1938). (Zbl. Math. **18**, 356.)

APPELL, P.: [1] Sur une classe de polynomes. Ann. Sci. Ecole Norm. Sup. (2) **9**, 119—144 (1880).

BATEMAN, H.: [1] The polynomial of Mittag-Leffler. Proc. Nat. Acad. Sci. U.S.A. **26**, 491—496 (1940).

BIEBERBACH, L.: [1] Lehrbuch der Funktionentheorie, vol. 2, 2. edit. Leipzig u. Berlin: B. G. Teubner 1931. — [2] Analytische Fortsetzung. Erg. Math., N.S. **1955**, Nr. 3.

BOAS, R. P., JR.: [1] Representation of functions by Lidstone series. Duke Math. J. **10**, 239—245 (1943). — [2] Polynomial expansions of analytic functions. J. Indian Math. Soc. (2) **14**, 1—14 (1950). — [3] Entire functions. New York: Academic Press 1954. [4] On generalized averaging operators. Canad. J. Math. **10**, 122—126 (1958). (§20.)

—, and R. C. BUCK: [1] Polynomials defined by generating relations. Amer. Math. Monthly **63**, 626—632 (1956).

BOURBAKI, N.: [1] Eléments de mathématique. XII. Fonctions d'une variable réelle, Chap. IV—VII. Actual Sci. Ind., No. 1132. Paris: Hermann 1951.

BOURION, G.: [1] Sur la relation de récurrence de Faber. Bull. Sci. Math. (2) **80**, 73—76 (1956).

BRENKE, W. C.: [1] On generating functions of polynomial systems. Amer. Math. Monthly **52**, 297—301 (1945).

BUCK, R. C.: [1] A class of entire functions. Duke Math. J **13**, 541—559 (1946). — [2] Interpolation and uniqueness of entire functions. Proc. Nat. Acad. Sci. U.S.A. **33**, 288—292 (1947). — [3] Interpolation series. Trans. Amer. Math. Soc. **64**, 283—298 (1948). — [4] Expansion theorems for analytic functions. I. Lectures on functions of a complex variable, edited by W. KAPLAN, pp. 409 to 419. Ann Arbor: University of Michigan Press 1955. — [5] On n-point expansions of entire functions. Proc. Amer. Math. Soc. **6**, 793—796 (1955).

BURCHNALL, J. L.: [1] The Bessel polynomials. Canad. J. Math. **3**, 62—68 (1951).

CARLITZ, L.: [1] A note on the multiplication formulas for the Bernoulli and Euler polynomials. Proc. Amer. Math. Soc. **4**, 184—188 (1953).

CARMICHAEL, R. D.: [1] Functions of exponential type. Bull. Amer. Math. Soc. **40**, 241—261 (1934). (§§ 9, 20.)

CARTWRIGHT, M. L.: [1] Integral functions. Cambridge Tracts in Mathematics and Mathematical Physics, No. 44. Cambridge 1956.

COOKE, R. G.: [1] Infinite matrices and sequence spaces. London: Macmillan 1950.

DeMAR, R.: [1] Existence of interpolating functions of exponential type. Trans. Amer. Math. Soc. **105** 359—371 (1962). — [2] Vanishing central differences. Proc. Amer. Math. Soc. **14**, 64—67 (1963). — [3] On a theorem concerning the existence of interpolating functions. Trans. Amer. Math. Soc. (to appear). — [4] A uniqueness theorem. Proc. Amer. Math. Soc. (to appear).

DENISYUK, I. M.: [1] Some properties of polynomials analogous to Laguerre polynomials. Dopovidi Akad. Nauk Ukrain. RSR. **1954**, 79—81. — [2] Some integrals and expansions which contain normalized Laguerre polynomials and their analogues. Dopovidi Akad. Nauk Ukrain. RSR. **1954**, 165—167. — [3] Some integrals, matrices and approximations connected with polynomials analogous to the Laguerre polynomials. Dopovidi Akad. Nauk Ukrain. RSR. **1954**, 239—242. — [4] Some relations which contain the normalized Laguerre polynomials and analogues to them. Dopovidi Akad. Nauk Ukrain. RSR. **1954**, 324—326. — [5] New polynomials analogous to the Laguerre polynomials. Dopovidi Akad. Nauk Ukrain. RSR. **1954**, 327—330. — These papers are in Ukrainian. Cf. Math. Reviews **16**, 694; **17**, 39.

DIENES, P.: [1] The Taylor series. An introduction to the theory of functions of a complex variable. Oxford 1931.

ERDÉLYI, A. (editor): [1] Higher transcendental functions. New York-Toronto-London: McGraw-Hill. Vols. 1, 2, 1953; vol. 3, 1955.

EVGRAFOV, M. A.: [1] Interpolyatsionnaya zadacha Abelya-Goncharova. Moscow: Gos. Izd. Tehn.-Teoret. Lit. 1954. — [2] The method of near systems in the space of analytic functions and its application to interpolation. Trudy Moskov. Mat. Obščestva **5**, 89—201 (1956) [Russian].

EWEIDA, M. T.: [1] Order of magnitude of the zeros of polynomials in basic series. Duke Math. J. **14**, 865—875 (1947).

FABER, G.: [1] Über polynomische Entwicklungen. Math. Ann. **57**, 389—408 (1903).

FALGAS, M : [1] Sur certaines fonctions associées aux bases de polynomes et leur utilisation à la définition des séries de base et à l'étude de l'effectivité de ces bases. I. La définition des séries de base. II. L'effectivité des bases de polynomes. C. R. Acad. Sci. Paris **242**, 1563—1566, 1677—1679 (1956). — [2] Sur les domaines étoilés vérifiant certaines propriétés liées a la propriété de convexité. C. R. Acad. Sci. Paris **244**, 2275—2278 (1957). — [3] Sur les séries de base relatives à certaines classes de fonctions entières. C. R. Acad. Sci. Paris **245**, 1208—1211 (1957). — [4] Sur la définition des séries de base de polynômes. C. R. Acad. Sci. Paris **249**, 2705—2707 (1959); **250**, 43—45 (1960); **252**, 2363—2365, 2493—2495 (1961). — [5] Sur l'effectivité des séries de base de polynômes. C. R. Acad. Sci. Paris **254**, 3296—3298, 3473—3475 (1962).

FASENMYER, Sister MARY CELINE: [1] Some generalized hypergeometric polynomials. Bull. Amer. Math. Soc. **53**, 806—812 (1947).

HARDY, G. H.: [1] Divergent series. Oxford 1949.

HILLE, E.: [1] Contributions to the theory of Hermitian series. II. The representation problem. Trans. Amer. Math. Soc. **47**, 80—94 (1940).

HUFF, W. N.: [1] The type of the polynomials generated by $f(xt)\,\varphi(t)$. Duke Math. J. **14**, 1091—1104 (1947).

—, and E. D. RAINVILLE: [1] On the Sheffer A-type of polynomials generated by $\varphi(t)\,f(xt)$. Proc. Amer. Math. Soc. **3**, 296—299 (1952).

JORDAN, C.: [1] Calculus of finite differences. 2d edit. New York: Chelsea 1947.

KRALL, H. L., and O. FRINK: [1] A new class of orthogonal polynomials: The Bessel polynomials. Trans. Amer. Math. Soc. **65**, 100—115 (1949).

LEVIN, B. YA.: [1] Raspredelenie korneĭ tselyh funktsii. Moscow 1956.

LOHIN, I. F.: [1] On an interpolation problem for entire functions. Mat. Sbornik, N.S. **35** (77), 223—230 (1954) [Russian]. — [2] On the representation of analytic functions by Faber polynomials. Mat. Sbornik, N.S. **36** (78), 441—444 (1955) [Russian].

MACINTYRE, A. J.: [1] LAPLACE's transformation and integral functions. Proc. London Math. Soc. (2) **45**, 1—20 (1938). — [2] Interpolation series for integral functions of exponential type. Trans. Amer. Math. Soc. **76**, 1—13 (1954).

MACINTYRE, A. J., and S. S. MACINTYRE: [1] Theorems on the convergence and asymptotic validity of ABEL's series. Proc. Roy. Soc. Edinburgh, Sect. A 63, 222—231 (1952).

MARTIN, W. T.: [1] On expansions in terms of a certain general class of functions. Amer. J. Math. 58, 407—420 (1936).

MIKSA, F. L.: [1] A table of Stirling numbers of the second kind, and of exponential numbers. Math. Teacher 49, 128—133 (1956).

NACHBIN, L.: [1] An extension of the notion of integral functions of the finite exponential type. Anais Acad. Brasil. Ciencias 16, 143—147 (1944).

NASSIF, M.: [1] On the mode of increase of simple sets of polynomials of given zeros. Proc. Math. Phys. Soc. Egypt 4, No. 4 (1952) 29—36 (1953).

NEWNS, W. F.: [1] On the representation of analytic functions by infinite series. Philos. Trans. Roy. Soc. Lond., Ser. A 245, 429—468 (1953).

OBRECHKOFF, N.: [1] Sur le développement des fonctions analytiques suivant des polynomes orthogonaux. Dokl. Bolgar. Akad. Nauk (C.R. Acad. Bulgare Sci.) 7, No. 2, 5—8 (1954).

PALAS, F. J.: [1] The polynomials generated by $f(t) \exp(p(x) u(t))$. Thesis, University of Oklahoma, 1955. [Dissertation Abstracts, Ann Arbor, Mich., vol. 16, 1146 (1956).]

PETERS, G. O.: [1] Boole polynomials of higher and negative orders. Bull. Amer. Math. Soc. 62, 7 (1956). — [2] Boole polynomials and numbers of the second kind. Bull. Amer. Math. Soc. 62, 387 (1956).

POLLARD, H.: [1] Representation of an analytic function by a Laguerre series. Ann. of Math. (2) 48, 358—365 (1947).

PÓLYA, G.: [1] Untersuchungen über Lücken und Singularitäten von Potenzreihen. Math. Z. 29, 549—640 (1929).

—, u. G. SZEGÖ: [1] Aufgaben und Lehrsätze aus der Analysis. Berlin: Springer 1925.

PORITSKY, H.: [1] On certain polynomial and other approximations to analytic functions. Trans. Amer. Math. Soc. 34, 274—331 (1932).

PRINGSHEIM, A.: [1] Zur Geschichte des Taylorschen Lehrsatzes. Bibliotheca math. (3) 1, 433—479 (1900).

RAINVILLE, E. D.: [1] Certain generating functions and associated polynomials. Amer. Math. Monthly 52, 239—250 (1945). — [2] Notes on Legendre polynomials. Bull. Amer. Math. Soc. 51, 268—271 (1945). — [3] Generating functions for Bessel and related polynomials. Canad. J. Math. 5, 104—106 (1953).

SCHIFFER, M.: [1] Faber polynomials in the theory of univalent functions. Bull. Amer. Math. Soc. 54, 503—517 (1948).

SCHOENBERG, I. J.: [1] On certain two-point expansions of integral functions of exponential type. Bull. Amer. Math. Soc. 42, 284—288 (1936).

SELBERG, A.: [1] Über ganzwertige ganze transcendente Funktionen. Arch. Math. Naturvid. 44, 45—52 (1941).

SEWELL, W. E.: [1] Jackson summation of the Faber development. Bull. Amer. Math. Soc. 45, 187—189 (1939).

SHARMA, A., and A. M. CHAK: [1] The basic analogue of a class of polynomials. Riv. Mat. Univ. Parma 5 (1954), 325—337 (1955).

SHASTRI, N. A.: [1] On ANGELESCU's polynomial $\pi_n(x)$. Proc. Indian Acad. Sci., Sect. A 11, 312—317 (1940). — [2] Some results involving ANGELESCU's polynomial $\pi_n(x)$. Proc. Indian Acad. Sci. Sect. A 12, 73—82 (1940).

SHEFFER, I. M.: [1] Expansions in generalized Appell polynomials and a class of related linear functional equations. Trans. Amer. Math. Soc. 31, 261—280 (1929). — [2] Concerning Appell sets and associated linear functional equations. Duke Math. J. 3, 593—609 (1937). — [3] Some properties of polynomial sets of type zero. Duke Math. J. **5**, 590—622 (1939). — [4] Some applications of certain polynomial classes. Bull. Amer. Math. Soc. **47**, 885—898 (1941). — [5] Note on Appell polynomials. Bull. Amer. Math. Soc. **51**, 739—744 (1945).

SINGH, VIKRAMADITYA: [1] Appell set of polynomials. Proc. Nat. Inst. Sci. India **20**, 341—347 (1954). (§§6, 9, 10, xi.)

SMITH, R. C. T.: [1] Generating functions of Appell form for the classical orthogonal polynomials. Proc. Amer. Math. Soc. **7**, 636—641 (1956).

STEFFENSEN, J. F.: [1] The poweroid, an extension of the mathematical notion of power. Acta math. **73**, 333—366 (1941). — [2] On the polynomials $R_\nu^{[\lambda]}(x)$, $N_\nu^{[\lambda]}(x)$ and $M_\nu^{[\lambda]}(x)$. Acta math. **78**, 291—314 (1946).

STEINBERG, J.: [1] Application de la théorie des suites d'Appell à une classe d'équations intégrales. Bull. Res. Council Israel, Sect. F 7 F, 55—68 (1957/58).

SUMNER, D. B.: [1] A generalized averaging operator. Canad. J. Math. **8**, 437—446 (1956).

SUYETIN, P. K.: [1] On the uniqueness of series of Faber polynomials. Dokl. Akad. Nauk SSSR. **105**, 423—425 (1955) [Russian].

SZEGÖ, G.: [1] Orthogonal polynomials. Amer. Math. Soc. Colloquium Publ., vol. 23, New York 1939.

TOSCANO, L.: [1] Una classe di polinomi della matematica attuariale. Riv. Mat. Univ. Parma **1**, 459—470 (1950).

TOUCHARD, J.: [1] Nombres exponentiels et nombres de Bernoulli. Canad. J. Math. **8**, 305—320 (1956).

UDAVOV, R., and R. OLEN: [1] Solution of the inverse problem of basic series. Uč. Zap. Pinsk. Ped. In-ta im. Burbak. **3**, 14—16 (1954) [Byelorussian].

WHITTAKER, J. M.: [1] Interpolatory function theory. Cambridge Tracts in Mathematics and Mathematical Physics, No. 33, 1935. — [2] Sur les séries de base de polynomes quelconques. Paris: Gauthier-Villars 1949.

WIDDER, D. V.: [1] The Laplace transform. Princeton: University Press 1941.

WILLIAMS, G. T.: [1] Numbers generated by the function e^{e^x-1}. Amer. Math. Monthly **52**, 323—327 (1945).

Index

Abel interpolation series 38, 66
Actuarial polynomials 42
ADHOC (Cf. PONDICZERY) 28, 41
Admissible path 10
ALPER 60
AL-SALAM 71
Angelescu polynomials 41
Appell polynomials 18 ff., 26, 28 ff., 49, 53, 61, 67 ff.

Basic hypergeometric function 54
— series V, 1 ff., 17, 23 ff.
— set 1
Basis 1 ff., 45
Bateman Manuscript Project V, 29, 33
— polynomials 43
Bernoulli polynomials 29 ff., 66
Bessel polynomials 43, 44
BIEBERBACH V, 12, 49
Binomial polynomials 34
BOAS 13, 18, 33, 45, 52, 61, 64, 65
Boole polynomials 31, 37, 38
Borel summability 12, 13
— transform 6, 8 ff., 48
BOURBAKI 28, 57
— polynomials 11, 57
BOURION 59
Brenke polynomials 19, 21, 26, 32, 45, 51 ff.
BUCK 13, 18, 33, 34, 38, 45, 52, 66, 71
BURCHNALL 42, 43

CARLSON's theorem 65, 70
CARMICHAEL 71
CAUCHY VI
— integral formula 4, 5
CHAK 45
Chebyshev polynomials 58, 60
Comparison function 6 ff., 21, 48
Completely monotonic sequence 55
Conjugate indicator diagram 9
COOKE 13
Critical type 28

DeMAR 37, 66
Denisyuk polynomials 41
DIENES 13, 49
Difference equations 70

Difference polynomials 34, 44, 54
Differential equations 67 ff.
DOSS 2

Effective in Ω 48
ERDÉLYI V, 29, 33
E-summability 12
Euler-Maclaurin summation formula 29 ff., 67
Euler polynomials 30, 70
EVGRAFOV 4
EWEIDA 2
— polynomials 46
Expansion class 3
— formula 3, 23, 48
Exponential numbers 42
— series, partial sums of 31

Faber polynomials 59, 60
FALGAS VI, 2, 3, 4, 48, 59
FINE 55
FRINK 43
Functional equations 67 ff.

Gegenbauer polynomials 58
GELFOND 66
Generalized Appell polynomials 17 ff.
Generating function (relation) 18 ff.

Hadamard multiplication theorem 49
HARDY 53
Hausdorff moments 9, 50, 55, 60
Hermite polynomials 31, 65
— —, reversed 45, 46
— —, squared 41
HILLE 32, 65
HUFF 18
Humbert polynomials 58
Hypergeometric function 42, 50
— polynomials 43, 44, 60
Hypotrochoids 59

Indicator 8
— set 9 ff.
Interpolation problem 11, 22

Jacobi polynomials 60
JORDAN 37

Kernel expansion method VI, 10 ff.
KRALL 43

Laguerre polynomials 16, 31, 40, 65
— —, reversed 32, 45
Laplace transform 6, 28, 62
Legendre polynomials 32, 46
Lerch polynomials 57
LEVIN 8
Lidstone series 13 ff., 30, 46
LOHIN 4, 60

MACINTYRE 4, 9
MAKAR 2
MARTIN VI, 28
MIKHAIL 2
MIKSA 42
Mittag-Leffler polynomials 38
— star 13
— summability 12, 13, 24
MOTZKIN 45
Multilateral Laplace transform 62
Multiple expansions 1, 3, 17, 22 ff., 54, 59
MURSI 2

NACHBIN's theorem 6
Narumi polynomials 37
NASSIF 2
— polynomials 46
NEWNS 2
Newton polynomials, interpolation formula 34 ff., 57, 66

OBRECHKOFF 43
Order, of polynomial set 2
—, of entire function 2, 9

PALAS 18
Partial sums of exponential series 31
— — of power series 28
— —, (C, λ) 53
Periodic polynomial set 46 ff.
— entire function 66
PETERS 31, 37
Pidduck polynomials 38
PINCHERLE 58
Poisson-Charlier polynomials 37
POLLARD 65
PÓLYA 6, 8, 27, 34
— representation 4, 8, 12, 47, 68
PONDICZERY. See ADHOC
Poweroids 33

PRINGSHEIM VI
Product star 12

q-difference polynomials 44 ff., 53, 54

RAINVILLE 18, 32, 43
— polynomials 46
— —, reversed 32
Representation of zero 25 ff., 54, 59, 67
Reverse of a polynomial 28, 32

SCHIFFER 59
SCHOENBERG 16
Selberg polynomials 34
Semibase 3
SEWELL 60
SHARMA 45
SHASTRI 41
SHEFFER 31, 33, 64, 67, 70
— polynomials 19, 33 ff., 64
SINGH 74
SMITH 61
Squared Hermite polynomials 41
Star 12, 13, 24
STEFFENSEN 33
STEINBERG VI
Stirling interpolation series 34 ff.
Summability 12 ff., 24
SUMNER 70
Supporting function of a convex set 8
SUYETIN 60
SZEGÖ 34, 58

TANTAOUI 2
TAYLOR's theorem VI, 11, 57
TOSCANO 42
Touchard polynomials 42
Trigonometric sums 16
TRUESDELL V
Type of an entire function 6
Type of a set of polynomials:
— à la Bateman Project 33
— à la SHEFFER 33
— à la WHITTAKER 2

Uniqueness theorems 65 ff.

WEISNER 38
WHITTAKER V, 1, 48, 61
WIDDER 55
WILLIAMS 42

Zero type (SHEFFER, of a set of polynomials) 33

Symbols

$A(D)$ differential operator 67, 69
$\mathfrak{A}(\Omega)$ space of functions analytic in Ω
   1, 48
$\mathfrak{B}_g$ 26
$B(\zeta)$ 21
$B_n(z)$ Bernoulli polynomials 29
$C(\varrho)$ 56
$C_n^{(\lambda)}$ Gegenbauer polynomials 58
$D = d/dz$ 67
$D(f)$ singularity set for Borel transform
   of f 7
e_q, E_q 45, 54
$E(\sigma)$, $E(\sigma, T)$ expansion class 2, 3
$_pF_q$ hypergeometric function 42
$G(\varrho)$ 57
$h(\theta, f)$ growth function 8
H_n Hermite polynomials 31, 45
I_0, J_0, J_λ Bessel functions 32, 43, 46
$k(\varphi, S)$ supporting function of S 8
$\mathfrak{R}_\psi$ all functions of finite Ψ-type 6
$K(z, w)$ representation kernel 4
$\mathscr{L}$ linear functional 3, 16
$L_n^{(\lambda)}$ Laguerre polynomials 16, 31, 32
$p^*(z)$ reverse of polynomial p 28
$\mathfrak{P}$ linear space of polynomials 1
P_n Legendre polynomials 32
$P_n^{(\alpha, \beta)}$ Jacobi polynomials 60
R_n Rainville polynomials 32

$\mathfrak{S}$ space of sequences 3
T, U linear transformations 3
$V(f)$ 48
$W(\zeta)$ inverse of $g(w)$ 21
Δ, V difference operators 35, 55, 70
Δ_w, Δ_ζ 21
Λ domain of regularity of Ψ 49
Ξ 52
Ψ 6, 48
Ω 1, 48
$\Omega^\bullet$ 48
$\Omega^\oplus$ 49
Ω_0 48, 52, 57
Ω_w, Ω_ζ, Ω_w^*, Ω_ζ^* 21, 24
$(E) \text{-} \sum a_n$ (E-summability) 12
$(ML) \text{-} \sum a_n$ (Mittag-Leffler summability)
   13
$(\gamma)_n = \gamma(\gamma + 1) \ldots (\gamma + n - 1)$ 42
S' complement of set S
$S_1 \cup S_2$, $S_1 \cap S_2$ union and intersection
   of sets
$S_1 \cdot S_2$ all $z_1 z_2$ for $z_j \in S_j$
$1/S$ all $1/z$ for $z \in S$
$S_1/S_2 = S_1 \cdot (1/S_2)$
$S_1 \odot S_2 = (S_1' \cdot S_2')'$
S^Δ convex cover (hull) of S
$\times$ Cartesian product of sets